Defining Values for Research and Technology

The University's Changing Role

Edited by
William T. Greenough,
Philip J. McConnaughay,
and Jay P. Kesan

ROWMAN & LITTLEFIELD PUBLISHERS, INC.
Lanham • Boulder • New York • Toronto • Plymouth, UK

ROWMAN & LITTLEFIELD PUBLISHERS, INC.

Published in the United States of America
by Rowman & Littlefield Publishers, Inc.
A wholly owned subsidiary of The Rowman & Littlefield Publishing Group, Inc.
4501 Forbes Boulevard, Suite 200, Lanham, Maryland 20706
www.rowmanlittlefield.com

Estover Road
Plymouth PL6 7PL
United Kingdom

British Library Cataloguing in Publication Information Available

Library of Congress Cataloging-in-Publication Data

Defining values for research and technology : the university's changing role / edited
by William T. Greenough, Philip J. McConnaughay, and Jay P. Kesan.
    p. cm.
  Includes index.
  ISBN-13: 978-0-7425-5025-4 (cloth : alk. paper)
  ISBN-10: 0-7425-5025-7 (cloth : alk. paper)
  ISBN-13: 978-0-7425-5026-1 (pbk. : alk. paper)
  ISBN-10: 0-7425-5026-5 (pbk. : alk. paper)
  1. Academic-industrial collaboration—United States. 2. Education, Higher—
Economic aspects—United States. 3. Universities and colleges—United States.
4. Research—United States. I. Greenough, William T. II. McConnaughay,
Philip J. III. Kesan, Jay P.
  LC1085.2.D42 2007
  378'.103—dc22   2006018424

Printed in the United States of America

♾ ™ The paper used in this publication meets the minimum requirements of American National Standard for Information Sciences—Permanence of Paper for Printed
Library Materials, ANSI/NISO Z39.48-1992.

Dedication

This project was undertaken with the support and sponsorship of the Center for Advanced Study of the University of Illinois at Urbana-Champaign. Established in 1959 as a special unit of the Graduate College, the Center for Advanced Study is charged with promoting the highest levels of cross-disciplinary scholarship and discourse. The center's professors include the late double Nobel Laureate John Bardeen (physics and chemistry); Nobel Laureates Paul C. Lauterbur (physiology) and Anthony J. Leggett (physics); National Medal of Technology recipient Nick Holonyak; Crafoord Prize in Biosciences recipient Carl Woese, and many other leading scholars who are professors at the University of Illinois. Information about the center is available at www.cas.uiuc.edu.

William T. Greenough
　Director, Center for Advanced Study
　Swanlund Professor of Neuroscience
Philip J. McConnaughay
　Dean and Farage Professor of Law
　Penn State University
Jay P. Kesan
　Professor of Law
　University of Illinois College of Law

Contents

Figures and Tables

FIGURES

TABLES

Introduction

Philip J. McConnaughay, William T. Greenough, Jay P. Kesan

This is a book about the values that should help guide the research mission of major universities. This topic is especially important during an era when government financial support of university research is in decline, causing universities and their professors to increasingly look toward the private sector, forming partnerships with profit-seeking corporations to support their research endeavors. Our focus is research-intensive universities: those major research universities where, as described by former Stanford president Gerhard Casper, "the intensity of research is part and parcel of the traditional university function of teaching and learning."[1] These are the universities whose discoveries, inventions, and intellectual property in biotechnology, nanotechnology, pharmaceuticals, computer software, engineering, and other academic disciplines increasingly drive our national and global economies while offering the possibility of great wealth to inventors, authors, and the universities at which they are employed.

We embarked on this inquiry with a notion of research universities as places where inquiries are made and research agendas pursued because of their intellectual challenge and potential importance to humanity. This somewhat idealized notion views research universities as places where knowledge and discoveries are shared resources, and advances are the product of criticism and contribution by a community of scholars. As Isaac Newton is said to have remarked, "If I have seen farther, it is because I have been standing on the shoulders of giants."

The assumption of research propelled by disinterested academic inquiry does not by any measure presuppose research universities to be "ivory towers" separated from the rest of society. To the contrary, the connection between research universities and the rest of society is very real and very deep.

"The great American contribution to higher education," in fact, as Sir Eric Ashby has noted, "has been to dismantle the walls around campus."[2] The establishment of land-grant universities,[3] the consequent pursuit of agricultural and industrial arts as academic disciplines, the opening of higher education to individuals from all segments of society,[4] and the increased use of public funds as a principal source of university sustenance and research[5] have combined to result in university research agendas deeply reflective of actual societal needs, problems, ambitions, and goals.

Until relatively recently, this deep connection to the external world did not seem particularly threatening to the academic values traditionally characteristic of research universities. Academic freedom and disinterested inquiry, peer review of applications for research funding, and the exposure of research data and conclusions to peer criticism and contribution remained characteristic of major university research activities. Even the federal government's massive financial support of scientific inquiry during the cold war came to be viewed over time as that of a "benign patron" and not as the government seeking to displace disinterested scientific inquiry in favor of preconceived outcomes.[6]

This perception began to change, however, during the 1980s and 1990s as a result of fundamental changes in the level and mechanisms of governmental support for university research, and in the relationship between research universities and private industry. With respect to the former, there has been a significant reduction during this period in governmental financial support of research universities. At public universities, the decline of state support has been dramatic, whether in absolute terms, as a percentage of university budgets, or as a percentage of state revenues.[7] A common lament among public research university presidents during this period is that their institutions have gone from "state *supported*" to "state *assisted*" to "state *related*" to "state *located*." One recent headline bemoaned, "The *Disappearing* State in Public Higher Education."[8]

At the federal level, even though the reduction of financial support has not been as significant,[9] it has been accompanied by a troublesome trend away from peer-reviewed research grants by the National Science Foundation (NSF), the National Institutes of Health (NIH), and other federal agencies toward "pork-barrel" earmarked grants to universities based not on the academic merit of research proposals, but on political maneuvering and lobbying.[10] This trend has been exacerbated by the end of the cold war. As the late U.S. representative George E. Brown Jr. observed in an essay, "Defining Values for Research and Technology,"[11] from which this book takes its title, "With the cold war behind us, we have no clearly defined values for research and technology sufficiently visionary to justify public confidence and support that scientists and engineers seek for their efforts."

At the same time, the 1980s and 1990s saw major new global industries emerge directly from basic research in information technologies, biogenetics, and biotechnology. The line between "basic" and "applied" research became increasingly irrelevant. Moreover, federal policy changed during this period, permitting universities to patent, license, and profit from their federally funded inventions and discoveries.[12] This has both promoted partnerships between research universities and private industry and whetted the appetites of researchers and universities alike to share in the enormous wealth being generated by the commercialization of the intellectual property they developed.[13] Research universities increasingly came to be viewed as engines of economic growth.

The coupling of these trends—the reduction in peer-reviewed governmental support of university research and the emergence of new intellectual property and economic incentives for university-industry research partnerships—has resulted in a "considerable reorientation of university research programs away from public (and largely open) research projects toward private (and more frequently proprietary) research projects."[14] This reorientation, in turn, has created a new set of challenges to traditional academic values—challenges that include both risks and opportunities.

The risks are most evident:

> Unfortunately, for some time now universities have placed more value on patents that bring in revenue than those that might show more originality. Thus inventing is viewed mainly as technology transfer, not as something with academic value of its own.[15]

> Once patents or profits become motivation for their work, they'll find it very hard to sort out what their research priorities are going to be.[16]

> Private companies and universities are busy patenting plant genes and claiming intellectual property rights to biotechnology advances. Suddenly, the free exchange of plant resources is in question as discoveries representing millions of dollars in profits are patented, potentially keeping poor farmers from using them.[17]

In 1998, the University of California at Berkeley's Plant and Microbial Biology Department struck a five-year deal with the Swiss biotech giant Novartis for $5 million in annual research support in exchange for Novartis gaining a right of first refusal to license or patent inventions emerging from Novartis-funded research.[18] Many saw this as an unsavory alliance more likely to skew academic research toward Novartis's business objectives than it was to support independent scholarly endeavors. In 2003, Columbia University sought additional patents on a soon-to-expire 1983 patented technique for producing biotech drugs from genetically engineered animal cells. Leading biotechnology companies sued the University, accusing it of illegally seeking to extend the life of the patent in order to sustain a revenue stream that had brought the institution hundreds of millions of dollars.[19] Instances such as these have caused former Harvard president Derek Bok to worry that "earlier [research university] norms about limiting the use of exclusive patent licenses, lending research materials, refusing to invest in faculty-sponsored business, and avoiding secrecy have all been eroded in the hope of making money."[20]

At the same time, the deepening relationship between research universities and private industry has revitalized major universities and scientific research in important ways. Recent years have seen more than one headline declaring, "Across the U.S., Universities are Fueling High-Tech Booms."[21] Industrial research parks now adjoin and work closely with major research universities in Palo Alto, Austin, Chapel Hill, Boston, San Diego, and too many other "university towns" to mention. As a recent article in *The Economist* noted, "It is the ability of American universities to spawn patents and entrepreneurs, rather than to collect Nobel prizes, that excites onlookers the most."[22]

A recent national report on university-industry synergies explained the phenomenon in this way:

> Regional clusters succeed because they are able to innovate more rapidly. They facilitate access to information, specialized skills, and business support. . . . Their proximity to universities . . . helps refine the research agenda, train new talent, and enable faster deployment of knowledge.[23]

Others have noted that deepening university-industry relationships simply reflect one aspect of the academic trend toward interdisciplinarity, which recognizes that many of the most important new thrusts in knowledge have emerged at the interfaces of previously separate academic disciplines.[24]

The fundamental question that emerges from the risks and opportunities of university-industry relationships is whether universities and their professors will pursue industrial imperatives at the expense of traditional academic values, or whether they instead will achieve new levels of innovation and create new industries in the pursuit of traditional academic values. Depending on the answer, university-industry partnerships could be of "great benefit to industrial expansion as well as enrich academic life."[25]

The values to which research universities and professors should adhere in managing the risks and opportunities of university-industry relationships is the core inquiry of this book. How can universities ensure intellectually vigorous and public-spirited research agendas when the temptation of profit seeking looms so large? How can universities ensure the open exchange of ideas and data when the commercial success of their inventions and industrial partners depends on secrecy or delays in disclosure?[26] How can universities license their intellectual property in ways that do not restrict use of the inventions for humanitarian purposes?[27] How can universities ensure the ongoing support and vibrancy of the humanities and other important academic disciplines unable to compete with the commercial promise of the sciences and engineering?

We asked some of the world's leading policy makers, scientists, and scholars to share their reflections on these questions. World-renowned developmental economist Lord Meghnad Desai suggests in his powerful essay that "universities made a mistake in becoming useful." Former National Science Foundation director Erich Bloch and National Center for Supercomputing Applications founder Dr. Larry Smarr suggest in their essays that universities have not become useful enough, and that partnering with industry will only increase their social and economic contributions. The father of the "Green Revolution," World Food Prize winner M. S. Swaminathan, provides an essay that reveals a life of science, innovation, and service not dependent on a university setting. These outstanding contributors are joined by eleven others of similar stature.

We have divided their responsive essays into four sections: (1) "The Purpose of the Research University"; (2) "Forging Partnerships with Industry and Governments"; (3) "Funding, Economic Incentives, and the Research Agenda"; and (4) "The Dark Side of University-Corporate Partnerships."

Noted physicist and chancellor emeritus of the University System of Maryland, Donald Langenberg, opens section I with a brief historical account of the research university and a positive prognosis for what he views as the university's natural progression toward ever-deepening partnerships with other sectors of society. Former advisor to the president on sciences and technology Jack Gibbons continues with an important reminder of some of the serious public questions societies face and an expression of concern about whether these questions will be addressed if short-term rather than long-term incentives begin to control university research agendas. Lord Meghnad Desai of the London School of Economics closes this section with a provocative look at core university functions that are not replicated by other institutions, and by challenging us to consider whether research universities might be better off severing their ties with industry and government and returning to their original purpose of pursuing the advancement of knowledge for its own sake.

National Medal of Science recipient and former NSF director Erich Bloch opens section II with a look at what he calls "the U.S. innovation system" and the important role of research universities as partners of government and private industry in that system. Founding director of the National Center for Supercomputing Applications Larry Smarr continues with an exploration of new models of research partnerships and with a provocative prognosis of "industry and education . . . interwoven at all levels." Former International Maize and Wheat Improvement Center (CIMMYT) director General Timothy Reeves then provides a powerful reminder, illustrated by the work of CIMMYT, of the role of basic research in creating public goods, but suggests that partnerships with industry and an acknowledgement of private intellectual property rights may be a necessary means of achieving that end. M. S. Swaminathan, revered father of the "Green Revolution" and World Food Prize winner, closes this section with another look at the creation of public goods and the role of research universities as "integral agents of change in society."

Former NASA chief scientist Kathie Olsen opens section III with an insider assessment of federal science and technology funding policy, a major driver of university research agendas. University of Virginia political scientist James Savage continues with a powerful indictment of federal earmarking and its effect on university research agendas, combined with an eloquent plea for the restoration and strengthening of peer review. University of Michigan law professor Rebecca Eisenberg closes with a comparative look at how the different incentives affecting university and private research pursuits interplay at times to promote the free public dissemination of private-sector discoveries and the intellectual property protection of research university discoveries.

Our book concludes with a look at "The Dark Side of University-Corporate Partnerships." Consumer advocate Michael Hansen opens this section with a highly critical account of what he views as the tendency of university-corporate relations to skew research agendas in favor of short-term profitable goals and away from longer-term agendas in the public interest that may not involve profit. University of California, Riverside, sociology professor Toby Miller then describes a corporatized university in which research institutions have become, in his view, "landlords, tax havens, and R&D surrogates rolled into one, with the administrators and fund-raisers lording it over the faculty." University of California, San Diego, comparative literature professor Maso Miyoshi closes with a lament that "higher education as a whole is currently experiencing a nearly complete loss of its historic purpose" with a growing divide between faculty engaged in commercially valuable pursuits and those who are not threatening the future of research universities as we know them.

Research universities today confront a potentially transformative challenge as they work to define the values that will cause their relationships with industry and government to enrich their research mission rather than detract from it. We hope that these essays contribute to that effort.

NOTES

1. Gerhard Casper, "The Advantages of the Research-Intensive University: The University of the 21st Century," address at Peking University Centennial, Beijing, People's Republic of China, May 3, 1998, www.stanford.edu/dept/pres-provost/speeches/980503.html.

2. As quoted by Derek Bok in *Beyond the Ivory Tower* (Cambridge, MA: Harvard University Press, 1982), 65. Sir Eric Ashby is a former vice chancellor of Cambridge University and master of Clare College.

3. Morrill Land Grant Act of 1862, act of July 2, 1862, ch. 130, 12 Stat. 503, 7 U.S.C. 301 et seq.; and Morrill Land Grant Act of 1890, act of August 30, 1890, ch. 841, 26 Stat. 417, 7 U.S.C. 322 et seq.

4. Two major historical events in this regard, in addition to the establishment of public land-grant universities, were the enactment of the GI Bill of 1944 (U.S. Statutes at Large, 1944, vol. 58, part 1, 288–94) (entitling armed forces veterans to publicly funded higher education), and the end to segregated higher education in the United States during the 1960s and 1970s.

5. Vannevar Bush's renowned call for government funding of scientific research in the aftermath of World War II, *Science: The Endless Frontier* (Washington, DC: Government Printing Office, 1945), coupled with the onset of the cold war, launched a postwar period of NSF, NIH, and other government funding of university research that is responsible for many of the most important scientific and economic advances of the last half century.

6. Roger Geiger, *Research and Relevant Knowledge* (New York: Oxford University Press, 1993).

7. See, for example, Sheila Slaughter and Larry L. Leslie, *Academic Capitalism: Politics, Policies and the Entrepreneurial University* (Baltimore, MD: Johns Hopkins University Press, 1997); *The Economist*, June 2, 2001, 29; Linda Cohen and Roger Noll, "Universities, Constituencies and the Role of the State," in *Challenges to Research Universities*, ed. Roger Noll (Washington, DC: Brookings Institution Press, 1998).

8. Jeffrey Selingo, in *Chronicle of Higher Education*, February 23, 2003, A22 (italics added).

9. See, for example, Steve Lohr, "Panel to Urge Big U.S. Effort in Technology," *New York Times*, February 20, 1999 ("Federal investment in most areas of information technology research has been flat or declining for nearly a decade, adjusted for inflation, largely because of the push to curb Government spending in general and because the end of the cold war brought a contraction in military-related research").

10. See, for example, Irwin Feller, "Research Subverted by Academic Greed," *Chronicle of Higher Education*, January 16, 2004, B6; and James P. Savage, *Funding Science in America* (Cambridge, UK: Cambridge University Press, 1999). See also William J. Broad, "Science Group Says U.S. Budget Plan Would Harm Research," *New York Times*, April 23, 2004.

11. George E. Brown Jr., "Defining Values for Research and Technology," *Chronicle of Higher Education*, July 10, 1998.

12. Beginning in the late 1970s, NIH policy changed to permit universities to secure patents on discoveries resulting from NIH-funded research. The Department of Defense and NASA followed suit. See Harvey Brooks, "Current Criticisms of Research Universities," in *The Research University in a Time of Discontent*, ed. Jonathan Cole, Elinor Barba, and Stephen Groubard (Baltimore, MD: Johns Hopkins University Press, 1998), 247. The Bayh-Dole Patent and Trademark Amendments Act of 1980 followed soon thereafter, expressly permitting universities to patent, license, and profit from inventions and discoveries arising from federally funded research.

13. The number of major research-university-held patents increased by more than 1,500 percent between 1984 and 1997, from 408 to over 6,000. James Duderstadt, *A Land-Grant Act for the 21st Century*, Prism Online, May–June 2000, www.asec.org/prism/may00/html/lastword.cfm.

14. Cohen and Noll, "Universities, Constituencies and the Role of States," 40.

15. Erich E. Kunhardt, "Necessity as the Mother of Tenure?" *New York Times*, December 14, 2004, op-ed.

16. Anthony DePalma, "The 'Slippery Slope' of Patenting Farmers' Crops," *New York Times*, May, 24, 2000, A2.

17. DePalma, "The 'Slippery Slope' of Patenting Farmers' Crops."

18. William Brand, "UC-Syngenta Collaboration up in the Air," *Oakland Tribune*, December 20, 2002; Goldie Blumenstyk, "Berkeley Joins Swiss Company in Controversial Technology-Transfer Pact," *Chronicle of Higher Education*, December 4, 1998.

19. See, for example, Andrew Pollack, "3 More Biotech Companies Sue Columbia over Patent," *New York Times*, July 16, 2003, A19; and "Schools Profit from Publicly Funded Research," Associated Press, at CNN.com, April 19, 2003, www.cnn.com/2003/education/04/29/patent.universities.ap/index.html.

20. Derek Bok, *Universities in the Marketplace: The Commercialization of Higher Education* (Princeton, NJ: Princeton University Press, 2003), 120.

21. Carey Goldberg, in *New York Times*, October 8, 1999, A20.

22. *The Economist*, "Universities," October 14, 1997, 17.

23. "Sustaining America's Prosperity," National Innovation Summit, April 5–6, 2001. See also Annalee Saxenian, *Regional Advantage: Culture and Competition in Silicon Valley and Route 128* (Cambridge, MA: Harvard University Press 1996), 2–3: "Silicon Valley has a regional network-based industrial system that promotes collective learning . . . among specialist producers of a complex of related technologies. The region's dense social networks and open labor markets encourage experimentation and entrepreneurship. . . . The functional boundaries within firms are porous . . . as are the boundaries between firms and local institutions such as universities."

24. Harvey Brooks, "Current Criticisms of Research Universities," in *The Research University in a Time of Discontent* (Baltimore, MD: Johns Hopkins University Press, 1994), 232.

25. Quoting Erich E. Kunhardt, "Necessity as the Mother of Tenure?" *New York Times*, December 14, 2004, op-ed.

26. See, for example, *The Economist*, "Scientific Publishing: Who Pays the Piper?" February 12, 2005, 78–79, reporting that, "On February 3rd America's National Institutes of Health (NIH), the world's biggest sponsor of medical research, announced that from May it will expect research work which it has helped to finance to be made available on-line, to all comers, and free, within a year of that research having been published in a journal." A countervailing influence since the September 11, 2001, terrorist attacks has been whether the open sharing of scientific information and data might be susceptible to misuse. See, for example, "Publish and Perish?" *The Chronicle of Higher Education*, October 11, 2002, A16; Diana Jean Schemo, "After 9/11, Universities Are Destroying Biological Agents," *New York Times*, December 16, 2002; and Nicholas Wade, "Science Panel Urges Review of Research Terrorists Could Use," *New York Times*, October 9, 2003, A1. But consider also Claudia Dreifus, "The Chilling of American Science" (a conversation with Nobel Laureate Robert C. Richardson), *New York Times*, July 6, 2004, Science Times, 2, quoting Dr. Richardson as saying, "One can't help but think that some of the post-9/11 restrictions are not motivated by security concerns, but by a misguided, almost 19th century protectionism of intellectual

property. There are policy makers who believe, 'We've got to keep all these foreigners from stealing our secrets and beating the pants off us.'"

27. See, for example, Andrew Pollack, "Universities to Share Patented Work on Crops," *New York Times*, July 11, 2003, reporting that, "Saying the development of crops that could feed millions of people is being choked off by biotechnology patents held by large corporations, several leading universities are joining to share information on their patented technologies and make them more widely available. . . . The universities say they will not let one another or other groups [license] their patented technologies [too] broadly, [preferring instead to reserve] rights to the technologies for . . . humanitarian purposes instead of giving total control to a single company."

Section I

The Purpose of the Research University

Donald Langenberg opens this section with a brief historical account of the research university and a positive prognosis for what he views as the university's natural progression with ever-deepening partnerships with other sectors of society. Jack Gibbons continues with an important reminder of some of the serious public questions societies face and an expression of concern about whether these questions will be addressed if short-term rather than long-term incentives begin to control university research agendas. Lord Meghnad Desai closes this section with a provocative look at core university functions that are not replicated by other institutions and by challenging us to consider whether research universities might be better off severing their ties with industry and government and returning to their original purpose of pursuing the advancement of knowledge for its own sake.

1

Research Universities in the Third Millennium: Genius with Character

Donald N. Langenberg

It was Goethe who penned the words, "Genius develops in quiet places, character out in the full current of human life."[1] Although that captures the essence of what I would like to discuss, being the typical academic, I cannot resist the temptation to expound on the profundity and insightfulness of Goethe's words for our topic at hand. Let us take, for example, Duke Ellington (Edward Kennedy Ellington [1899–1974]). Ellington was a prodigious composer. Jazz aficionados consider him America's own Bach. He composed his music on train rides, subways, and even in the bathtub. He composed in the waning, lonely moments before drifting off to sleep. He scribbled musical notes on restaurant napkins and shirtsleeves. Ellington's genius found its expression in the quietest, most contemplative moments when he was alone.

There is a certain similarity in the way a researcher works. The scientist's work is often slow, frustrating, and even plodding at times, similar to Ellington's struggle to find the perfect note or musical phrase. Then comes one of those eureka-type moments, the moment when an idea or concept suddenly comes together, as if it had been there the entire time.

Genius creates wondrous things. Yet genius alone is not enough. Just as the consummation of the creation of music is when it is heard by its listeners, knowledge, with all its consequences, must be widely communicated and accessible in order to have lasting value. These things do not happen in isolation, but rather in the context of the "full current of human life." For both individuals and institutions, their external interactions simply bring to the fore their fundamental values and shape their reputations. These values and reputations form the foundation of what Goethe then called "character."

For individuals and institutions, and, more relevantly, for universities, both genius and character are extremely important because there needs to be the innovation that comes from genius and the dissemination of that knowledge that is determined by its character. The question, then, is how do American universities, as a whole, measure up?

If Nobel Prizes are indicative of genius, universities are doing very well. Regarding character, defined here as the values that the university is driven by, together with its reputation, the university as an institution ranks high on the list.[2] However, there is some indication that we cannot afford to be complacent about the public's high regard for the university. The university therefore needs to continuously and faithfully uphold its commitment to the public good as well as the production of genius. The theme of this book's inquiry astutely reflects a widespread sense of unease, both toward the university and other spheres of the public domain. This is because the role of the university is changing, and whether it is for the better or worse is the topic of much discussion. Of course, the dynamism of the university is not a new phenomenon, although during some time periods, change came at a pace that makes continental drift seem swift.

THE NATURE OF THE UNIVERSITY
(PAST, PRESENT, AND FUTURE)

A Brief Look Back

Going back to the earliest universities in Europe, the first universities were essentially vocational schools, established to train young men for learned professions such as the clergy, law, and medicine. Then, during the Enlightenment, the idea of the university as the home for the pursuit of "knowledge for knowledge's sake" was born. Thus, students focused their studies on subject matters like the classics, philosophy, mathematics, and the like. For some, it meant allegiance to the notion of "pure thought" or "pure mathematics," meaning that the knowledge produced was somehow unsullied by any earthly practical value. It was here that concepts like academic freedom and institutional autonomy, ideas that eventually formed the core values of the university, arose. Thus, the "ivory tower" was born.

It was in the nineteenth century that the creation of two institutions radically impacted the university as it was, creating a conflict within the ivory tower that continues today. The first was the creation of the German research university, based on the proposition that the principal function of the university should be the creation of new knowledge, a departure from the university's theology/seminary function until that time. The second was the American invention of the land-grant university. This incorporated another element into the function of the university, the direct engagement of new knowledge and practical application, especially in agriculture and the mechanic arts. And although higher education has undergone many different changes since the late nineteenth century, these key roles of the research universities in the life of our nation have remained the same. During the cold war, for example, university research and technically trained graduates contributed invaluably to our national security and earned strong public support. Following the end of the cold war, university research and the university's skilled graduates have become essential to the social vitality and economic development of the nation.

The University Today

This increasing functional importance of the university has resulted in an explosion of partnerships between research universities and the world around them. Government and business leaders are demanding greater engagement than in the past, and along with that comes a myriad of pressures on institutions of higher education. The question is, is the university making Faustian bargains that will radically transform the university as it has been?

For most universities, the answer is probably yes. However, again, radical change itself is not new to the university. But whether the change will be for the better or worse will largely depend on the university community. Here is where the character of the university will be rigorously tested, out in the "full current of human life."

The Future of the University

Some critics urge a move back to the ivory-tower ideal and a rigorous fortification of the barriers between the university and business/industry. They yearn for the days when higher education operated more autonomously, when there was less concern about the applications of knowledge and more focus on its creation. Yet many would argue that this distinction between "pure" and "applied" is a false dichotomy, much like that of "teaching" and "research." Learning is learning, regardless of how it is accomplished, whether in a lecture hall or a research laboratory, and almost all learning eventually becomes applicable to some real-world situation.

The perspective or attitude one adopts does affect the relationship between the university and corporate America. Those who favor partnering with industry applaud the university's responsiveness to real-world concerns and trends. Those who criticize such dealings say that some of us are doing too much to shape our programs to industry's patterns and placing too much emphasis on training students for jobs instead of providing them with a broad-based liberal education.

If one listens to the most vociferous critics, the tone may sound as if it is the high-minded and noble "us" versus the crass and greedy "them." One might wonder whether proponents of such laments would adhere so strongly to their principles if offered multimillion-dollar gifts from big-name companies and private firms for scholarships, new laboratories, and endowed professorships. Probably not, provided that there were no strings attached. But the reality is that there is "no such thing as a free lunch," and there are always strings attached. Donors are rarely concerned with simply making a magnanimous gift. Inevitably it boils down to judgments about donor's intention(s), that is, whether the strings attached are consistent with the university's mission and goals. This process filters through the values of the university. Adherence to the core values of the university falls into Goethe's realm of character.

There are those who are concerned that partnerships between universities and industry will cause universities to become reduced to "mere" trade schools and that the growing emphasis on technology transfer will cause them to become more concerned with profits and bottom lines than with teaching and learning. Stanley Ikkenberry, president emeritus of the University of Illinois, succinctly summed up that view when he wrote, "The danger is that you unwittingly lose sight of the ultimate aim of education and that you confuse commercialization with the deeper purposes of higher learning."

First, the environment for technology transfer from university labs to industry drastically changed with the passage of the Bayh-Dole Act in 1980. Prior to Bayh-Dole, it was federal policy that any inventions produced as a result of federally funded university research belonged in the public domain. At first glance, this policy seems fair. The public should own what it pays for. However, the problem lies in the fact that if everybody owns a patent, then nobody is likely to make the very large investment required to turn that product into a commercially viable product. As a result, prior to 1980, the federal government had thousands of unexploited patents in essence collecting dust on its shelves, so to speak, with no benefit to the public.

The Bayh-Dole Act transferred ownership of inventions arising from federally funded research to the institutions actually performing the research. The effect on research universities and their relationships with industry was dramatic. Before 1981, fewer than 250 patents were issued to universities each year.[3] During the 1999 fiscal year, universities filed for 7,612 patents in the United States and many more overseas. The Association for University Technology Managers (AUTM) estimates that the technology transfer activities of universities, teaching hospitals, and research institutes have added more than $40 billion to the U.S. economy and have supported 270,000 jobs. Alan Greenspan attributed at least 70 percent of the economic boom of the 1990s to technological advances, many of them driven by our research universities. Even our loftiest universities have recognized that there is no shame in teaming up with business and industry to advance research or develop commercial products. A case in point is Columbia University, which has done more than perhaps any other institution to aggressively commodify its intellectual capital—that is, the knowledge, research, and teaching of its professors.

Though this may seem like a novel idea, it is actually a modern version of what the Morrill Acts of 1862 and 1890[4] directed our land-grant universities to do for the nation's major businesses, agriculture, and the mechanic arts. The Morrill Acts were passed in order to allow federal funding to be allocated for the dissemination of knowledge to the public and also to do research that would have direct and practical application.

As for the issue of "reducing" the university to a mere trade school, one could argue that the university, from its inception, has always been a type of trade school. The most ancient and honored European universities began as places to teach men how to be bishops, judges, and physicians. Academics are fond of distinguishing the "learned" professions from "mere" trades, or "education" (noble but useless) from "training" (lowly but useful). Such distinctions are becoming irrelevant today. Most jobs require a mixture of education and skills training. No one doubts that the surgeon who might remove a tumor from your brain needs both a deep understanding of current research in the neurosciences and the technical skill to remove the tumor without paralyzing you. As one educator is fond of saying, in the future there will be two kinds of people, "educated and unemployable."[5]

If we accept that our universities have the obligation and challenge of educating and training people who can perform at very high levels in future jobs and occupations, then do we not have an obligation to attempt to determine whether we are actually successfully accomplishing this? This is one of the main reasons for maintaining close relationships with all organizations and enterprises that employ our graduates. We can then learn whether we are doing right by the graduates, either by asking the graduates themselves or their employees.

In a study initiated by the Pew Charitable Trusts "At Cross Purposes: What the Experiences of Today's Doctoral Students Reveal about Doctoral Education,"[6] about four thousand doctoral students across eleven disciplines within the liberal arts and sciences at twenty-seven different universities responded to a questionnaire. The summary findings included the following:

- The common sentiment is that doctoral students are not receiving what they want, and their training is not preparing them for the jobs they take.
- Many students do not clearly understand what doctoral study entails, how the process works, and how to navigate it effectively.

It is difficult to avoid the conclusion that the gentlest criticism one could make about doctoral education in U.S. research universities is that it is dysfunctional. This is noteworthy because this problem lies right at the heart of what defines the research university, that is, to produce trained, technically skilled professional researchers. The question is, how will we respond to this situation? Again, we see the necessity and need for character. These examples take us back to the wisdom of Goethe that "the greatest genius will never be worth much if he pretends to draw exclusively from his own resources."[7]

CONCLUDING THOUGHTS

This new world, in which research universities find themselves more and more tightly linked with the outside world, poses many challenges. Some are brand new, while others are old friends in new guises. Intellectual-property and conflict-of-interest issues are but two of the vexing issues at hand. Our universities are at a crossroads, where, on the one hand, they could choose to retire from the fray and wall themselves up in their ivory towers. On the other hand, they could plunge into "the full current of human life" and make the best of what they encounter there. The character and loyalty of the university will be sorely tested. Clark Kerr, president emeritus of the University of California, said,

> The danger is not that loyalties are divided today but that they might be undivided tomorrow. . . . I would urge each individual to avoid total involvement in any organization; to seek to whatever extent lies within his power to limit each group to the minimum control necessary for performance of essential functions, to struggle against the effort to absorb.

Let me briefly list a few issues that will test the mettle of the character of the future research university:

- The IT revolutions in which our universities are already leading actors, where intellectual-property and conflict-of-interest issues are prominent, along with the implications of the IT revolution for our own enterprise.
- The national crisis of performance in our schools. Teachers are the key to resolving this crisis, and we in higher education are responsible and obliged to improve both the quantity and quality of our nation's teachers.
- Intercollegiate athletics—a headline in *The Chronicle of Higher Education* plaintively asked, "Can anyone do anything about college sports?" The Knight Commission, composed of some of higher education's most distinguished leaders, has been grappling with this question for a decade, without noticeable result. This is one area that needs both closer examination and follow-through.

- Concerning doctoral education, it does not seem to be meeting the needs of those being educated. Again, the question is, what will we do about it?

This incomplete list is daunting, if not downright intimidating. However, the situation is optimistic, because higher education and the research university have faced many challenges in the past, and oftentimes today's problems have become the launching pad for tomorrow's solutions. As we pursue the new opportunities that come with these challenges, it is possible that we will behave in ways that compromise our institutions' fundamental missions and values. How the university then responds will depend less on its genius and more on its character. Ultimately, it is our character that must preserve the genius that is, in fact, our academic soul.

NOTES

1. This quote could also be translated, "Talent is nurtured in solitude; character is formed in the stormy billows of the world." [Ger., *Es hildet ein talent sich in derStille / Sich ein Charakter in dem Strom der Welt.*] From *Torquato Tasso* (1, 2, 72).

2. According to a more recent Harris poll (2002, January 16–21), "Confidence/ Trust in Institutions" (this poll can be viewed at www.pollingreport.com/ insitut.htm), Americans ranked the military first, the White House second, and major educational institutions following fourth. The question was, "[A]s far as people in charge of running [see table 1.1] are concerned, would say you have a great deal of confidence, only some confidence, or hardly any confidence at all in them?" N = 1,011 adults nationwide.

Table 1.1.   Confidence and Trust in Institutions

A Great Deal of Confidence (%)

	1/02	1/01	1/00	1/99
The military	71	44	48	54
The White House	50	21	21	22
The U.S. Supreme Court	41	35	34	42
Major educational institutions, such as colleges and universities	33	35	36	37
The executive branch of the federal government	33	20	18	17
Medicine	29	32	44	39
Television news	24	24	20	23
Organized religion	23	25	26	27
Congress	22	18	15	12

3. The Bayh-Dole Act of December 12, 1980, Pub. L.No. 96–517, 94 Stat. 3015–28 (codified as amended at 35 U.S.C. 200–11, 301–7[2000]).

4. The Morrill Acts include the Morrill Act of 1862, 7 U.S.C. 301 et seq., and the Morrill Act of 1890, 7 U.S.C. 322 et seq.

5. Personal conversation.

6. Golde, C. M., and T. M. Dore, *At Cross Purposes: What the Experiences of Doctoral Students Reveal about Doctoral Education,* report prepared for the Pew Charitable Trusts, 2001, at www.phd-survey.org.

7. This frequently cited quote by Johann Goethe can be found at www.cybernation.com/quotationcenter/quoteshow.php?id=38085.

2

The University of the Twenty-first Century: Artifact, Sea Anchor, or Pathfinder?

John H. Gibbons

Now that we have traversed across the ceremonial millennium line, it is important to think ahead. Of course, trying to look at a whole century is a little further than most of us can see. But thinking about the century as a whole is necessary because we want to gain insight into the future and from that gain a better sense of the role of the university and its products, namely the production of knowledge and the training of educated people. C. P. Snow (1905–1980), the famous English author and physicist, once said, "A sense of the future is behind all good politics. Without it one can leave nothing, either decent or wise, to the world." But sometimes this type of anticipation of the future is discounted because of the feeling that the future seems irrelevant amidst the urgency of the moment.

Nonetheless, anticipation is important. Albert Schweitzer said, "Mankind has lost its capacity to foresee and therefore to forestall. We will end up destroying the earth."[1] This is a tragic statement coming from a man who spent so much of his life working for the betterment of humanity. The time for foresight, along with the resulting ability to forestall, is very much upon us. This is particularly true in the academic and research community, since we are the ones that need to deal with things outside of the traditional time frame and marketplace of either business or politics.

In this chapter, I will first take a look at a few of the key issues we confront in the twenty-first century before going back to consider the university. We are constantly discovering the magnitude of our ignorance, and I would like to consider this potential in the context of our responsibilities as both stewards and consumers of the fruits of the earth.

SOME ISSUES OF THE TWENTY-FIRST CENTURY

Population Growth

World population more than tripled during the last century, exceeding six billion by the end of the century. Population growth is a powerful function with an enormous amount of momentum. This high momentum shows that we have to think about our responsibilities as the earth's stewards in order to enjoy the earth's fruits in the long term.

Figure 2.1 reflects the number of people using resources, together with the economic growth of consumption (using iron ore as a surrogate) per person. This chart indicates that consumption of resources—the product of population growth times economic growth—grows even faster.

Figure 2.2 shows that population growth in industrial countries is leveling out while Third World countries continue to undergo enormous expansion in population. Even though birthrates are in fact dropping in some parts of the developing world, the momentum of population will carry them to enormous population growth in this century.

Figure 2.2 presents an optimistic picture concerning the leveling-out global population. But it also implies a time of drastic demographic transition, which will be one of the greatest challenges of the twenty-first century. In the United States, we are facing a similar challenge today with the so-called baby

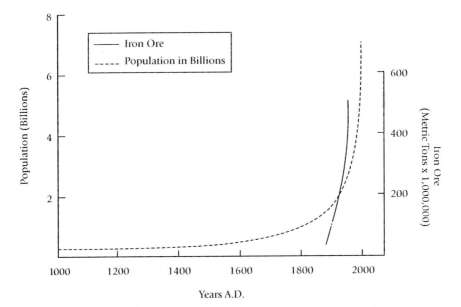

Figure 2.1.   The Exponential Rate of Increase of Consumption and Population

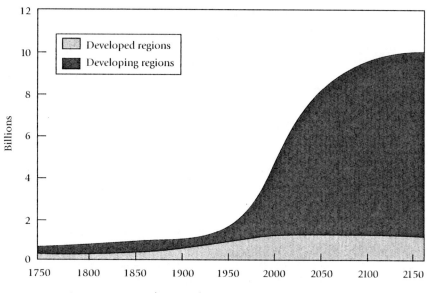

Figure 2.2. Population and Its Correlation with Development

boom bulge moving through the ranks of our society. With more people retiring than entering the workforce, the result is a steady depletion of Social Security trust funds and an increase in Medicare funds. However, this is nothing compared to other parts of the world that are more rapidly undergoing similar kinds of transition, particularly in China and other countries that are desperately trying to regulate their birth and death rates.

Population is a fundamental dynamic that does not yield easily to rapid change. Not only do people have a sense of being individual families but also of being part of a world of families. I recall a particularly telling cartoon depicting a boat packed with rabbits, with one rabbit saying to the others, "I'm only gonna say this one more time. Our only chance is self-control."

The twenty-first century is, in a sense, a "moment of truth" for global population dynamics and the move away from exponential growth toward a leveling off of the global population. This transition is going to be an increasingly critical issue for the entire world, not just for developing countries.

Resources and the Environment

Economic growth is typically about 3 percent of gross domestic product (GDP) per year. A 3 percent GDP growth compounds to about a twenty-three-year doubling time. When calculated over the course of a century, 3 percent growth per year means growth by a factor of eight in a century. If the population also doubles over the century, then the net impact on global resources is approximately a factor of sixteen.

In this new century, we have the challenge of rethinking the way we use materials in order to create goods and services; that is, we need to begin thinking in terms of *closed* rather than *open* systems, a process called green manufacturing. Although there are various terms for this, it is basically the challenge to the science and engineering communities to devise ways of creating and providing wealth, in terms of goods and services, with a much lower net flow of materials through the economy. This is the time to fret about how to straighten out our loss of stratospheric ozone. This problem alone will take us half a century before it is back under control. It is a time in which we might base our new economic production system on "dematerialization" rather than the old notion that "the more resources you consume, the better your GDP."

Another challenge to managing our resources is coming to grips with what have been called "nature's numbers"—that is, services provided by so-called "natural" ecosystems such as the supply and purification of fresh water, the distribution of atmospheric temperature, the pollination of plants, and other services that are provided by the biosphere in the natural environment. The value of these services has been estimated to be equivalent to the sum of all the other things that humans provide themselves through their own manufacturing capabilities. In other words, at least half of our total real wealth may be tied up in these ecosystem services. But they are not part of our national economic accounts. And when they are not part of the accounts, we fail to measure them as we progress, therefore, devaluing them. One of the great challenges will be capitalizing on work done by some outstanding researchers, such as Ken Arrow at Stanford,[2] who are trying to quantify the values of ecosystem services and get them into our national accounts so that we might have more rational decision making about their use.

Another part of our resources challenge concerns energy, along with its companion, global climate change. We know that the average surface temperatures of the earth and atmospheric CO_2 concentration tend to track one another, with surface temperatures increasing along with increases in atmospheric CO_2. The concern is that the present level of CO_2 in the atmosphere is now about 380 parts per million, which is higher than the earth has experienced over perhaps a million years or more.

CO_2 is important because in addition to water vapor and methane, it is part of the greenhouse cover that keeps infrared radiation from reradiating from the earth and upsetting the temperature balance between the input energy from the sun and its emission from the earth's surface. Deforestation, burning fossil fuels, and a variety of other human activities are driving us into a whole new era of earth's history. The issue is, where will this take us in the future and what can we do about it, if anything? This century is a century-long "moment of truth" for global climate change, just as it is for population control. Scientists are concerned that, if we double preindustrial CO_2 levels, an era of instability, warming, and damaging climate change awaits us.

There is, in fact, real concern about the damage and destruction this century caused by a number of different global weather phenomena such as the inundation of lands by the sea as it expands thermally, the severity of storms, and the location of monsoons. Suppose it is decided that we should not exceed CO_2 levels of 550 parts per million. If we accept that number and calculate what we need to do in terms of tons of carbon emitted per year, we are faced with several scenarios. Scenario one: we keep doing what we have been doing—in other words, business as usual. That scenario takes us beyond 1,000 parts per million by 2200. On the other hand, if the resulting global atmosphere never exceeds 550 parts per million, we find that CO_2 simply cannot be dropped that rapidly, because it would wreck our economies. So we are talking about a reasonably well-behaved path to enable us to end up at levels of 550 parts per million, even though we exceed that number along the way.

What does this imply? Human activities emit about five billion metric tons of carbon per year. Of this amount, the United States alone emits about five tons per person per year, amounting to approximately ten times the emission of a typical person in a developing country. If we look to the end of the century, 2100, the total emission will probably go through a peak, and it will be back down to about five or six tons of carbon emitted per person per year. In contrast, the global average in this case is only one ton per person per year. We have to get back to these levels by the end of this new century in order to hold CO_2 levels to twice preindustrial concentration.

This is indeed no small challenge when taking into consideration both economic and population growth. If CO_2 levels were uncontrollable, that would be one thing, *but there are many things that we can do about it*. As one example, we all know that if we improve the mileage of our transportation fleet—whether through the development of alternative fuels or otherwise—we can achieve major reductions in petroleum demand.

Obviously, there will be other environmental challenges this century, such as new pathogens, whether through mutations responding to a "less-than-propitious" use of antibiotics or through rerelease of existing but confined pathogens through deforestation. The health of the biosphere, and its people, is going to be a continuing issue, since the problem of infectious diseases has yet to be solved. These and other challenges are uniquely suited to university-based solutions.

CHALLENGES FACING THE UNIVERSITY

What about the research university, then? What are the implications for the university community? First of all, there is a growing trend toward greater interaction between the traditional disciplines and the developing new disciplines. For example, computer science emerged from the interaction between mathematics, statistics, and electrical engineering. Fields like climate change studies and other new kinds of amalgamations of sciences will be established to tackle, understand, and resolve these challenges. This approach calls for an integration of different fields in the sciences and engineering in order to achieve the kinds of advances that are needed.

The confluence of science, engineering, mathematics, and computer techniques has led us to the threshold of new opportunities in health, biology, and medicine. In turn, this development requires people who are trained in these different disciplines to understand each other's "languages" and to be able to work in teams to make these advances. Interdisciplinary units will continue to grow in importance for research universities.

Second, new transinstitutional ties and researcher-funding links will connect colleagues within the university with institutions and colleagues outside the university. Transinstitutional links are areas of great concern because links between industry and its resources can lead to shorter time perspectives in the kind of work that the university chooses. For example, if you are working with energy, then you are naturally drawn into a multidecade perspective. It typically takes half a century to transition from one form of energy to another as a dominant source. But in the political and industrial world, the time constant is either two years or until the next election. Adlai Stevenson Jr. said, "We never see the handwriting on the wall until our back is up against it." His was an insightful observation because that happens all the time, where we get jerked around by not allowing ourselves to think, plan, and act on the future.

A related development that I foresee is the increase in continuing education and lifelong learning programs. A person graduating from college fifty years ago might have looked forward to one or two different job changes in his or her career, but now that is closer to half a dozen or more, and this requires flexibility and constant learning. This fluidity also is reflected in the cycling between academia and public- and private-sector life. These multiyear experiences in fact serve to enrich the lives of these lifelong learners, as they give their time and expertise in government and in industry, also better equipping them to return to the university. This is an important factor to consider because it should and will be a feature of the changing university.

The funding relationship between the university, governments, and industry will also continue to change. We have moved from state-funded to state-assisted to state-affiliated, and now to state-located universities. Simultaneously, the enormous economic value of research universities in terms of building the state's future economy has encouraged partnerships with the private sector and a reexamination of the place of universities in state budget priorities. A presidential executive order under former president William Clinton was passed clarifying and standardizing the federal government's position on government-university partnerships across agencies.[3]

Finally, there is the issue of intellectual property. Pragmatically, support of exploration in science, higher education, and technology development has been a part of our national history since the drafting of the Constitution. The problem is in deciding between investments for public knowledge or personal profit. An individual firm with a capitalistic spirit can capture the rewards of that supported research for profit development. Thinking about the human genome mapping project and other biomedical and related science projects, it is much easier for the average person to feel connected with that type of work because of the connection to his or her health concerns. It is more difficult to think about how a superconducting supercollider somehow affects one personally. However, this is an evolving saga that is not immediately solvable, but one moving with us into this new century. It is a little unsettling, but in Thomas Jefferson's words, "Our laws and institutions must change with the times," and dealing with these issues is simply part of the process.

CONCLUDING THOUGHTS

Returning to my title, the twenty-first-century university is at once an artifact, sea anchor, and pathfinder. However, in understanding these terms, one must see how the three concepts are linked. First of all, these relationships are not mutually exclusive because an artifact brings with it the lessons and the meanings from the past and is important in contextualizing the present in terms of the past. James Madison wrote to Jefferson in 1789 that "the Federal business has proceeded with a mortifying tardiness. We are in a wilderness without a single footstep to guide us. Our successors will have an easier task." They were building on artifacts of history, of other civilizations, of other times. They masterfully took advantage of information from those artifacts in order to create the foundations of our present government. We want to hold on to the artifact feature of the university.

A sea anchor implies being more deliberate, testing ideas, and as the saying goes, "looking before we leap." There is a Native American saying, "Don't test deep water with both feet." A key feature of scientific research, or the scientific method, is just that—to be sea anchors testing, retesting, and challenging ideas, making them more sound. Dan Boorstin once said that the greatest obstacle to discovering the shape of the earth and the continents and the oceans was not ignorance but the *illusion of knowledge*. A sea anchor implies testing, evaluating, and making sure we are not under the spell of some kind of illusion. Sea anchors do not hold us back as much as they assure us as to where we are going. Finally, the pathfinder—pathfinding implies leading, discovering, and inventing, which have played and will continue to play key roles in our history.

The questions we should ask include the following: What do we mean by growth and progress in the new century? Do we think of it in terms of the expansionist model that we have successfully run with for a long time? Or do we take the exponential model as an anachronism and turn to people like Rene DuBose, who spoke about amenities that are peculiarly human, like open space, quiet spaces, things that feed us in a very important way but are not yet on our national economic agenda? We must keep in mind that there are ways of thinking about progress and growth that are different from traditional measures that we have been caught up with in more recent years.

The twenty-first century is going to be the time in which we look for a miracle while looking to ourselves and asking, what can we do that makes sense? What can we learn from the artifacts of the past, from the sea anchors of the present, and from the pathfinding of discovery and invention? In a sense, our task is what Saint-Exupéry said in *The Little Prince*: "Your task is not so much to *predict* the future but rather to *enable* it." I think that it is the role of the university to help us understand what is happening, to help chart a course and pick out the appropriate advances in human intellect, and to help guide society toward a much happier future. Like it or not, this new century is a "moment of truth" for us and for our universities.

Following are Dr. Gibbons's responses to questions posed during a discussion of his paper at the University of Illinois Center for Advanced Study.

Q: There does seem to be a synergy between the university community and the individual sector of our society. At least we can see where both sides profit and where the hazards are. I have trouble seeing the same type of synergy between the university and the federal government. I think that university people feel like they can contribute to governmental decisions. On the other hand, I don't think that the people in government recognize the value of the university. I wonder if you'd comment on whether or not the politics of our country wants to recognize the existence of the university other than in paying lip service to it.

A: *I think that thoughtful people in government do understand its essential role, and in fact the president's executive order[3] spends a lot of time looking at the relationship between the two. The driving principles of the executive order are: (1) research is an investment in the future, (2) the integration of research and education is vital, (3) excellence is promoted when investments are guided by merit review, and (4) research must ultimately be conducted with integrity. These principles reinforce the notion that our institutions of higher learning are the best places to conduct research because they have so many collateral benefits. The government also recognizes how much the government depends on the university community for a trained supply of educated people and also for the fruits of research. It's a symbiotic relation-*

ship that certainly is reflected in the federal government spending billions of dollars per year in basic research. The real question is how that partnership can be made better, but belief in the partnership is evident.

Q: The real question seems to be how that partnership can affect public policy?

A: *I think that maybe we could say that this is a problem of the government, but it is also a problem of the universities. I think an unspoken but increasing responsibility of the university community, especially in the science and engineering community, is to be more openly engaged in public dialogue and informing society, including elected officials, about what is going on in these areas and being able to offer or see options in dealing with them.*

Q: You mentioned scientists leaving the universities to work for the government and in private industry and then coming back to academia. Are you suggesting that universities should think about that and create such opportunities?

A: *I certainly am, and I might be prejudiced about this but I think that when you leave a particular profession for a while, go out and have a different experience that adds to your experience base, it stretches you in various directions and potentially makes you a more valuable person. Not necessarily to go back and do what you were doing, but perhaps to play a new role in the university. For example, I think that the Center for Advanced Study[1] is an excellent example of a center that is cross-disciplinary, working to create a forum for different experiences. It is, in fact, a wonderful mixing pot for the university and is an excellent example for the trend for universities in the twenty-first century.*

Q: Concerning the humanities, it would be an interesting thing to count out how many English/humanities majors have minors in the sciences.

A: *That is a wonderful idea, because we are, in fact, entering a time in which the sciences and engineering are increasingly pervasive in our personal and national lives, and we can ill afford to become illiterate in the areas that are so critical to our own governance. Madison said that we must equip ourselves with the power that knowledge gives us if we mean to be our own governors. If we move more and more in the technocratic sense in our society and farther and farther away from understanding what it is all about, then we will lose our capability to govern ourselves.*

Q: In your opinion, is the lack of multidisciplinary collaboration a limiting factor in the creation of new technology?

A: *I think that is the difference between bureaucracies and universities. In a bureaucracy, strength is frequently viewed as derived from sequestering information and using it very selectively. The great strength in the university is that it derives strength from sharing information. This free flow of information is an enormous source of strength in our society. It is a very precious asset and commodity that we would be making a bad mistake to let get away from us.*

NOTES

1. Ann C. Free, *Animals, Nature, and Albert Schweitzer* (Washington, DC: The Albert Schweitzer Fellowship, the Albert Schweitzer Center, the Animal Welfare Institute, and the Humane Society of the United States, 1982), www.awionline.org/schweitzer/as-idx.htm.

2. More information on Ken Arrow can be found at his website: www.econ.stanford.edu/faculty/arrow.html.

3. Referring to Executive order 13185, 66 FR 701, passed by former president William Clinton (January 3, 2001).

4. The Center for Advanced Study is an interdisciplinary unit at the University of Illinois, Urbana-Champaign.

3

Can Universities Survive the Global Knowledge Revolution?

Lord Meghnad Desai

After briefly introducing how universities came about, I will argue that universities are going to be nonviable in an increasingly global world, and, consequently, new institutions may need to be formed in order to fulfill the traditional functions of the university. It may be that some universities will clearly reinvent themselves and become this new vehicle suitable in our present world, but that remains to be seen. Clearly, universities are facing difficulties performing their traditional roles.

One notable peculiarity of the university is the longevity of its existence. Plato's Academy was probably its forerunner, and Aristotle is the archetype of the consummate interdisciplinary professor embodied in one person. Universities were founded centuries before the common era (CE) began. Among these universities are included the University of Takshashila (now Taksila) in Afghanistan, various Buddhist universities in India, the Nalanda University in the Bihar state (India), and others. Nalanda and Takshashila date from roughly the fourth century BCE and lasted about a millennium. During the Dark Ages and earlier in medieval Europe, universities began and flourished. The universities of medieval Europe set the model for subsequent universities that would be established. This particular university model has now spread everywhere, to America, Asia, and Africa. In the process, however, it has also undergone modifications, mostly due to geographical contextualization along with a constantly changing world.

AN OVERVIEW OF THE TIMELINE

In their early days, universities were essentially seminaries, training young men to take up the holy orders. Students were taught theology, spiritual disciplines, and related matters. In a broader sense, these men were assigned the task of learning something, like theology and philosophy, for the purpose of disseminating that information into the society at large. As these men grew older, they went on to become ministers, clergymen, statesmen, secretaries of state, chancellors, and in some cases, very prominent and pivotal in running kingdoms, such as Thomas Cromwell and Sir Thomas Moore. In a sense, the university performed the dual function of training these men as scholars and the clergy, and in the process, making them fit to run the affairs of state.

After the Reformation, the university became more secular, and the pursuit of "knowledge for knowledge's sake" emerged. For the first time, the university began training people not only for church or civil service, but also for other university careers as professional researchers and scholars. Science was no longer an amateur undertaking. The university became the center for science research in addition to its teaching function. Nonetheless, in Europe, amidst all these changes, education remained an elite privilege for a long time.

Higher Education in the United States

In the United States, however, this elitist tradition changed in the nineteenth century with the establishment of the land-grant university. This institution had a different philosophy about education. It was an education geared at the citizenship rather than for the elite. And though the establishment of these universities did not attract people en masse initially, the idea that education was relevant and useful in daily life for the average person began taking root.

The twentieth century represents the next major turning point for the university in the United States. The two world wars brought to the fore the fact that scientific and technological research was extremely vital to national security and spurred an enormous national investment in scientific research and development (R & D). In fact, World War II was a war more or less determined by its R & D component. Technology had to be invented in the universities and quickly engineered into production. The most striking and famous example of this was the construction of the atom bomb. The bomb was an international collaboration of physical scientists, and the splitting of the atom a joint enterprise of universities in both Europe and the United States.

The Post–World War II Years

After the conclusion of World War II in 1945, universities proliferated and expanded, in part because of the influx of students (in the United States, the

G.I. Bill provided an extraordinary incentive), and in part because the university became incorporated into national funding systems. Yet, with all these things taking place, universities had to maintain their commitment to pursuing knowledge with universal scope, knowledge that is independent of monetary incentives. The ideal picture of the university academic as a neutral, unattached scholar who serves knowledge was changing as the university became part of the state apparatus. In fact, however, the university traditionally had always been either attached to the state or to the church, serving whichever was the authority at the time. There has always been tension between university scholars who wanted to claim that they were neutral, independent, and incorruptible, on the one hand, and on the other hand, lucrative external pressure to serve the purposes of the state or church.

Nonetheless, the post–World War II era saw the university become part of what President Eisenhower called the "military-industrial complex" and contribute to the national economy both in the United States and in the UK and Europe as never before. In this capacity, universities prospered. The prestige of universities continued to grow, money situations became more favorable, student enrollment increased, and so on. Universities took on both a useful and respectable function. They were no longer simply arenas of critical thought; they were important parts of the state economic machine.

THE UNIVERSITY AND THE WORLD TODAY

Then two things happened. First, the phase of nation-based capitalism came to an end. Globalization has come hand in hand with a globalized national economy. The movement of capital, along with the growth of trade, has made national territorial capitalism a much weaker entity. People trade with other firms abroad, or firms trade with their own branches abroad; capital moves fast; and industry is footloose. In many ways, the power of the nation-state to shape its own economy has been compromised. Knowledge is becoming increasingly more important as an input to the global economic machine.

Up until this time, the university had functioned both as a gatekeeper and a producer of knowledge, certifying whether something was knowledge or not. It has been in response to this monopoly of knowledge production, however, that we are now seeing the emergence of alternate knowledge-producing organizations, organizations that today challenge the university as the primary knowledge producer.

Universities are an anomaly, because although they are considered nonprofit institutions, they do play a key role in bringing in revenue. For example, the economic value of the university does not necessarily accrue to itself but more likely is channeled to some favored faculty, siphoned off by businesses with contractual relationships with universities, or perhaps realized by the government. What is problematic about this is the way that universities produce and disseminate research, which may not, in fact, be carried out in the most economical way. The university's methodology is that of an old-fashioned, nonprofit, and anachronistic organization. Although the teaching at some undergraduate institutions closely resembles a factory-type mode, universities by and large are very labor-intensive, small-scale production outfits, which cannot realize the economies of scale. We still think that person-to-person contact is a good quality for a university to have, and in fact institutions like Oxford and Cambridge still use the personal tutorial as a mode of teaching.

Universities in the United States have traditionally produced knowledge through a skills-focused, workshop type of framework. But this knowledge was never marketed as a commodity. It has been part of the university ethos to not be a mercantile commercial organization. Knowledge was to be made available, free of charge, to anyone who was qualified to be at the door by their suitable degrees. The charged fees, then, are just a small proportion of the true economic cost of producing knowledge.

The past decade has seen universities discovering that the intellectual property they produce has a high market value. In some sense, universities should capture the market value of the intellectual property they produce, driving the university in a more market-driven direction. But of course the whole university cannot go in that direction, which is where the tension comes in. People are interested in the intellectual property of things like genomics, physical processes, and technology-related things, but not Shakespearean criticism, theology, or anthropology. If the university is going to impart knowledge universally, the dilemma is going to be in the reality that its commercially viable operations are not sustainable on the same basis as its non-commercially viable but important tasks.

And so, in some sense, by "spreading itself too thin," the university seems to be perpetuating this tension within itself. On the one hand, knowledge has become more of a commodity today than ever before, but on the other hand, people no longer accept the belief that knowledge produced by the university is necessarily the best-quality knowledge, or even that a person has authority because of his or her academic credentials. In fact, Thomas Kuhn argues in his landmark book, *The Structure of Scientific Revolution*, that scientists are, in fact, more restricted than furthered by the university in the search for truth.

Knowledge is thus commodified, and in the process, diversified. We do not have "one knowledge" but "diversity in knowledge," and some of that knowledge is being produced outside of the university. And that knowledge is just as useful as that which is produced in the universities.

IMPLICATIONS FOR THE UNIVERSITY

In a world like this, then, what can a university hold out as its own? What does a university in fact do, and what then can it claim? What is the quality claim that universities can make to survive? The university's claim to legitimacy is the idea that its knowledge is better, more durable, more testable, more reliable, and so forth.

One of the most important challenges facing the university is whether the production of knowledge, research, and scholarship, in fact, is compatible with dissemination of knowledge. Up until today, university academics passionately believed that teaching and research were interconnected. For the academic, research and teaching were both high priorities. While some academics do enjoy both teaching and doing research, increasingly top-notch scientists, economists, and others are attracted to research-based institutes where there is no teaching, just strictly research. If this gravitation toward pure research institutions is indicative of a trend, then it is a sign that the joint production of research and teaching need not actually be the most efficient way of generating knowledge.

We also need to distinguish how the "corporate university"—in other words, an educational unit within a corporation—is different from the traditional university. The whole principle of the university is that education is broad, and so you learn other things regardless of your field. By creating a more efficient means of teaching people specific skill sets "in house," private corporations have shown that one can create fragmented, partial teaching programs that concentrate narrowly on one class of subjects or skills without bothering with other, peripheral classes/activities that universities require. And these fragmented, more specialized institutions are demonstrating higher levels of efficiencies versus costs as a criterion.

Another area of the university that is under fire is its traditional role as an information gatekeeper. Originally, universities were repositories of information as well as knowledge. Later, with the advent of the printing press, some of that knowledge became more generally available. Now, with the possibility of downloading anything off the Internet, information is easily accessible. Information gathering no longer has to be done at the university, since one can access information from home computers.

In order to survive, then, the university must define its core competence. For example, the university must establish distinctions between information and knowledge. While information can be gathered by anybody, knowledge enables one to do something with it. Anybody can read the *Encyclopedia Britannica* either online or in print. However, it takes knowledge both to read the pages and to know that only one or two pages are necessary for the purpose at hand. More knowledge is essential for determining where to go next to link that information to other information, and in making the different bits of information cohesive so that it becomes valuable. Information by itself is not valuable, but when information is gathered into some cohesive unit, it becomes knowledge, and thereby valuable. In a world, then, where anyone can download anything, the university must distinguish itself by making the claim that actually attending university will enable you to turn that information into knowledge.

A second important point that universities need to establish is that imparting information or even skills is not an end in and of itself. When universities first started becoming important in Western Europe, they taught theology. The theology taught had little relation to the material reality in which people lived, since theology was unscientific. However, it did not prevent the theologically well-grounded person from running the world around him. This seems like the pivotal example of what universities can and should do.

Universities use particular subjects of more-or-less practical relevance as conduits for developing critical thinkers. This point adequately emphasizes the fact that universities exist to teach people how to think and think independently. Theology, accountancy, economics, or physics may be the vehicle, but in fact the subject at hand does not really matter. Students are allowed to read indiscriminately. It has no bearing on the choice of a subsequent career. The university produces people who can think critically.

Students are taught to challenge and argue with texts, to interact with their fellow students and teachers, and to arrive at defendable conclusions using a variety of tactics, including rhetoric and argumentation. Eventually, they have to write an essay, take a final examination, or write their dissertation. Notice that all these things, in fact, are exercises in independent thinking. Students should be able to construct an argument on their own, and in a manner by which the skeptical reader will be convinced that they are right.

This ability to think critically is a rare, but not a specialized, skill. Once one knows how to think, he or she can think in a variety of situations and solve various puzzles and problems. And the university is unique in the fact that there are no other specialized institutions that teach people how to think. British universities are especially proud of the fact that they are able to produce undergraduates who study philosophy, economics, classics, literature, and so forth and then go on to become civil servants or industrialists or professionals because they were able to think independently.

The niche that universities have, then, is that in their production of knowledge, they can also impart the ability to think critically. The priority, then, is not just the actual pursuit of the acquisition of knowledge but the reshaping, reexamining, and then critically processing and synthesizing of the information, either with a lack or an onslaught of information.

CONCLUSION

Thus, we arrive at two concluding points. Universities should not be in the business of simply imparting information, since information can be collected through media like the Internet. We need to distinguish between information and knowledge. Knowledge is information with attached value, and as it is processed, it becomes more acute, coherent, and connected.

Second, the current state of production of knowledge and its dissemination is both inefficient and uneconomical. Things like teaching hundreds and thousands of undergraduates will no longer be advantageous to the university. Those activities will probably be hived off, and if universities do invest in them, they will do them separately from their core functions. Because of the pressure to deliver the standard product via textbooks and examinations, teaching at the undergraduate level is becoming commodified, with no scope for deviation and an increasing similarity to a factory. And, again, universities are not advantaged in pursuing that particular operation because there are other organizations that can do that.

The university is then left with the production of knowledge and research. And to the extent that a small number of apprentices and students can join that research activity, they will become the critical, independent thinkers that society will need and will call upon. However, it takes a considerable amount of time to teach people how to think independently. Many of the factory-type undergraduate teaching institutions are unable to foster independent thinking because of the sheer numbers of students and the nature of such a large university. Nevertheless, there remains a need for critical thinkers. Thus, it is the university that is entrusted with this role of teaching people how to think. It is the university's responsibility to cultivate independent thinkers.

As I mentioned earlier, people no longer blindly accept the authority of academia. In fact, the market is replete with authoritative knowledge from which an individual can choose. But, again, it is my belief that universities have the historic authority in that they have established methodologies and procedures such as peer-group review, certain independent scrutiny of evidence, replication of experiments, and critical reviews by referees. Only after passing certain procedures, such as testing a product for health, safety, and a variety of possible influences, can a product be launched without much trouble. It still remains the case that reliable knowledge in natural sciences and technology, when produced outside of the university, has seldom passed the test of peer review. Things like "cold fusion" immediately come to mind. Different people have made a myriad of claims about their medicines and biological or physical research that have not been able to withstand the scrutiny of replication and peer review.

While Thomas Kuhn argues that scientific revolutions are taking place despite rather than because of universities, I retain the view that, by and large, these scientific revolutions are much exaggerated. What really generates knowledge is the day-to-day practice of peer review, seeing it pass through a number of tests, launching the product, and getting it approved. This particular production technology, the production of knowledge, was developed in the university. This particular production of knowledge is the greatest achievement of the university.

The technique of knowledge production through critical peer review has not been copied or improved in any other organization. Private corporations cannot do that, although they can do other things. When you set up certain national research laboratories, they have to follow the particular methodology involving the production of knowledge that peer-group review has established and is unbeatably the preserve of universities.

In my view, then, two things will aid the survival of the university in this age. First, the university will continue to be the principal vehicle for developing critical thinkers. Second, it will continue to be the producer of valuable knowledge by adhering to its historical process for knowledge production. Other things like the university's contribution to human capital accumulation, economic growth, and managerial thinking, though not unimportant, can be done better by agencies other than the university. Universities made the mistake of becoming useful. The university needs to return to being a specialized producer of knowledge via a particular kind of process, and by being rather old fashioned, that process has less fear of being made obsolete than many of the recent "technological fads" that many universities have got into.

While the state, church, and corporations will always try to beset universities for their purposes, universities need to get smaller (away from the factory university model) and tougher, and they need to concentrate on the production of knowledge and critical thinkers. If the university can do this, then it will survive. Otherwise, it will just become a nondescript, inefficient factory of resources that can be better produced inside corporations.

Following are Lord Desai's answers to questions posed during a teleconference discussion of his paper at the University of Illinois Center for Advanced Study.

Q: There has been a trend, and it may be nothing more than that, for companies to adopt a new type of management theory involving knowledge management, where businesses consider their principal assets to be not capital per se but what their workers and managers know, and to share what they know more clearly. Do you think that this trend, if it bears out, might create a role for some of these alternative institutions of teaching to develop people with critical thinking skills?

A: *One of the things that I was trying to say was, when corporations set up an in-house "university," they are very much aware that knowledge management is a major task of a corporation. And I do think that the growth of institutions like these takes away some of the functions of the university. But what I call "critical thinking," thinking which goes against the grain, thinking that is unconnected to the purpose at hand, is not required by the corporation. The corporation wants thinking about the problem at hand. There will be a need for directive thinking in corporations, but what they will not need is critical thinking, thinking which is partly subversive, which mates through a certain restructuring of old knowledge and old ways of thinking. That remains very much the university's task.*

Q: How would you select students who would be able to carry out critical thinking? And where is the money going to come from for this type of proposition? It seems that an endeavor/reform like this would be very expensive and labor intensive, requiring major overhaul of the university as it is currently.

A: *I think you are right. I see universities as being much smaller and more selective because, in fact, the more popular things like training undergraduates will be done elsewhere. Many of the tasks that are done now at the undergraduate level, in fact, can be done elsewhere and are completely unnecessary to the university. Today we select people through certain procedures at various levels, such as GREs, SATs, et cetera. I think those kinds of tests are actually very useful to weed out who will and will not be able to advance to the next stage of education. However, I am a great believer in the unseen examination, which takes the person and examines him or her in person. By and large, that method of selecting people has not failed. I think that the current acceptance procedures are not bad. What they will have to become, that is, transform into, is to be able to select those who genuinely want to get into the production of knowledge and independent thinking. Imagine a world, then, where universities have no undergraduate component of their education, just graduate parts for research and scholarship. The people who want to advance will be fewer, because knowledge—information acquisition—can be done elsewhere. But how to finance this poses a dilemma. Currently, universities are facing increasing difficulty in financing their students. Students rely on bank loans, endowments, and government grants, et cetera. I imagine we will have to have some sort of loan contract— people borrowing against a future type of income in order to gain this type of knowledge. And students will basically finance themselves from their own future incomes, which is essentially what is taking place now. But I quite agree that it is very labor intensive, that it will be like the Oxford, Cambridge universities, which are high-quality universities because the student-staff/teacher ratios are very low. That type of Plato's Academy interaction with teachers, critically thinking, with well-qualified students, might be all that is left. It is not an elitist model, because everything else we want to do can be done elsewhere, and so people can specialize through other means. But those few people who want to focus on getting into the production of knowledge and thinking would pay the money to be able to do this, if they can, because they might have even worked for a few years before coming back to attend the university.*

Q: Could you say a little more about the differences between this high-quality knowledge that the university would be producing and the other type of knowledge produced by corporations?

A: *I think basically corporations have a directed purpose. They only need that knowledge which is directly relevant for their projects or the task at hand. They are basically going to be narrowly focused on selecting that kind of knowledge. Universities are not in that kind of business, because the whole purpose of the university and the beauty of knowledge creation is that you do not know beforehand what is going to be useful. One of my favorite examples took place at London University about twenty years ago when we were under considerable pressure from the government to only focus on useful things. I saw the vice chancellor of London University, after he had come back from talking to the minister, and he quoted the minister, saying, "You should do useful things," and he told me that he agreed. I told him that he should never have agreed about this, because, I said, "medieval theology is not a useless subject."*

Medieval Islamic theology seemed like a useless subject until the Ayatollah Kohmeni came to power in Iran—and until the more recent terrorist acts in New York and Washington, D.C.—and then everybody was looking around for someone who knew about Islamic theology. Now, to know about Islamic theology and what mullahs do requires four years of undergraduate work, four years of graduate studies, and then extensive scholarship after that in order for someone to authoritatively be able to advise a government on what this man means when he says such and such. To interpret the sayings of a distinguished theologian like the Ayatollah, you would have to know quite a lot about doctrine, background, and so on. In 1979, there were very few people who could authoritatively advise a government on what was happening in Iran, or, for example, in 1989, when the Soviet Union broke down, there were very few people who had studied central Asia. One of the great things about universities, in fact, is that we do not know today what will be useful tomorrow. Universities have to have a stake in that kind of research because the university will cast a net wide, and they will be able to do things that are not a priori seen to be useful by corporations. The point is corporations don't need to do this, and they have no time to do this, to waste resources waiting for the next new surprise to turn up. Universities admit people exactly for that purpose. It is not actually research, and in fact it is really a search. This is a search and a discovery, rather than research, and to that extent it is different.

Q: There is a movement in the United States to try to develop these economic clusters where you would have universities interacting closely with corporations, and they bring different people and different skills to bear, working in a synergistic way and producing economic benefits for society beyond what they could do individually, that is, both from the corporate and the university side. What is your view on this type of economic cluster? How do you think it fits in with your idea of a university?

A: *My view on this is that universities have gone down that road in the postwar period, first of all with governments, and only later with corporations. This part of the university, while very profitable, can be easily copied by other institutions. I do not think that the university has a unique selling point in that type of synergistic interaction, because corporations will be able to clone that bit of the university in house or by having consultants off campus or things like that. So, yes, it is very popular these days, but I do not see how universities gain in any sustained fashion. As an example, I spent a lot of my early days teaching econometrics, which is a statistical modeling kind of economics, and I started writing a computer program, teaching people the theory of that. This took a lot of time. All these things can now be done using very user-friendly programs. A lot of that training has thus been devalued. You need them to interpret evidence, think, and make inferences, which is the purpose of university instruction, but all the mechanical things are reproducible elsewhere. So, my thinking is that someday those types of clusters are going to throw universities out and advance without them.*

Q: If that's the case, then do you think that universities should avoid those types of relationships in the first place and focus more on fundamental types of knowledge searching, not research, but search for knowledge and understanding?

A: *Yes, I think that if you cannot avoid those types of relationships, you should at least unpack those things and hive them off separately so that those kinds of enterprises are hived off, and university's core activities are done differently. One thing which takes place in these economic clusters is that certain subjects get badly neglected, such as anthropology, English, archeology, and the humanities in general. In a university setting, there should be a kind of equality among all forms of knowledge and equal possibility of useful discoveries in all forms of knowledge. For example, when the United States got involved in the Vietnam War, there were not three people who could speak, read, or write in Vietnamese in the entire United States. Where were all the linguists or language experts? Well, people did not think that knowing languages was a very practical thing, and so people did not bother to get language skills. Well, languages are a part of humanities. So while universities need to invest in lucrative endeavors, eventually the core activities and values need to be protected and other things hived off and run as independent activities.*

Section II

FORGING PARTNERSHIPS WITH INDUSTRY AND GOVERNMENTS

Erich Bloch opens this section with a look at what he calls "the U.S. innovation system" and the important role of research universities as partners of government and private industry in that system. Larry Smarr continues with an exploration of new models of research partnerships and ways of structuring interdisciplinary research, and with a provocative prognosis of "industry and education . . . interwoven at all levels." Timothy Reeves and Kelly A. Cassaday then provide a powerful reminder, illustrated by the work of the International Maize and Wheat Improvement Center (CIMMYT), of the role of basic research in creating public goods, but they suggest that partnerships with industry and an acknowledgement of private intellectual property rights may be a necessary means of achieving that end. M. S. Swaminathan, revered father of the "Green Revolution" and World Food Prize winner, closes this section with a similar look at the creation of public goods and the role of research universities as "integral agents of change in society."

4

The Changing Nature of Innovation in the United States

Erich Bloch

The Center for Advanced Studies (CAS) and the University of Illinois (UIUC) chose wisely when they embarked on an inquiry focusing on the changing role of the university. With the rapid changes in our economy, industry, and science and technology, the university is struggling to adapt in an altered environment.

To provide a baseline for the changes in universities, this chapter focuses on the transformation that has occurred and is occurring within the U.S. innovation system. Some people might wonder if there is such a thing as a U.S. innovation system, but I would argue that the collective might of the federal and state governments and academic institutions, as well as the increasing number of private-sector foundations, not-for-profit organizations, and professional technical associations, all form a formidable assembly of activities that exert tremendous influence on individuals, institutions, and the country as a whole.

If there is any doubt that we are amidst far-reaching changes in our innovation system, consider the fifteen years that ended the last century. In 1985, the U.S. standing in technology and leading products was ebbing low. At that time, Japan was its most prominent and prosperous competitor, leading with its innovative technology, such as semiconductors, autos, textiles, and manufacturing prowess in general, and possibly, by some standards, in computer technology as well. It was even unclear who held the keys to the future, as it seemed like the United States could be overtaken by Japan in these areas. Compounding the problem was the hard-hit U.S. economy, which was in recession, causing both inflation and unemployment to be at high points. And of course there was the ever-increasing budget deficit. In this dismal state of affairs, both industry and the federal government's capability to increase R & D investments were in jeopardy. With few exceptions, there was neither a willingness to cooperate and partner between sectors, like academia and industry, nor a willingness to cooperate and partner with other competitors within the same industry, even when it was evident that collaboration did not hinder competition in the marketplace.

By the year 2000, Japan was no longer our main competitor and was barely managing to keep its head above water. The U.S. economy had seen major advances in its GDP, with the lowest unemployment figures in decades, and with budget surpluses and low inflation as well. The same industries that were in decline in 1985 were among the industry leaders in 2000, and what is more, new developments like the Internet, biotechnology, and personal digital appliances suggested U.S. leadership into the future. Industry R & D was increasing at exponential rates, and where partnership was once the exception, it had become, in fact, modus vivendi.

The years since 2000 have seen dramatic reversals of these trends, with the bursting of the high-technology stock bubble and the disruption of the world economy following the terrorist attacks of September 11, 2001. But the economic rebound since those events has been significant, and the United States seems, once again, to enjoy the prospect of sustained economic growth.

FACTS AND VIEWS

The question, then, is why is the United States continuing to prosper? Economists will find many reasons. Surely there is no single cause that can explain as complex a system as the U.S. economy. But it is now well recognized that technology is a big contributor to the economy. Let me quote Chairman of the Federal Reserve Alan Greenspan's explanation of the 1990s as supplying an extended explanation: "An economy that twenty years ago seemed to have seen its better days is displaying a remarkable run of economic growth that appears to have its roots in ongoing advances in technology."

Put this together with Robert Solow's (Nobel Prize winner in economics) observation that over 50 percent of our GDP increases are due to technology, and there is ample reason for the federal government to focus on our innovation system—namely science and technology development—along with developing the infrastructure of the institutions that serve that system. We need some assurance that this system will perform in the future as well as it is performing today. Performance, however, is not a given, since the foundations for today's technological successes were laid one or two decades ago. That is the lead time we need to take into account.

On a microeconomic level, the results are interesting. Certainly the productivity increases during the last five years of the past decade (figure 4.1) have been greater than in the past for a similar time period. This result, in part, was due to our investment in R & D going back about twenty years, and partially because of our focus on defense, which was driven by cold war concerns.

This is demonstrably true with semiconductors and IT in general, more specifically with computers and communications. The productivity growth in the IT sector in the 1990s was 20 percent greater compared to the low of 1.5 percent for all the other industry sectors.[1]

But there are other forces at work: the last decade certainly was the decade of the entrepreneur, made possible by developments like the Internet, the development of new industry sectors like biotechnology, the availability of capital from the United States and Europe because of favorable economic conditions (figure 4.2), and the development of new opportunities in academia and industry partnerships.

This vibrant entrepreneurship resulted in a large increase in the number of firms, as can be seen in Figure 4.3.

This is especially evident in the pharmaceutical and software areas. The benefit to the public is clear. Our national income increased by 50 percent from 1985 to 2000, and increases in per capita GDP rose 35 percent over the same time period (figure 4.4).

Productivity Growth

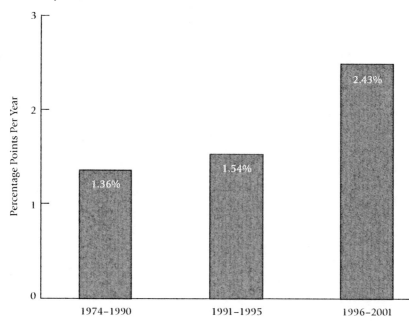

Figure 4.1.   U.S. Productivity Increasing

Total Net Stock of Private Fixed Assets
Trillions of Current U.S. Dollars

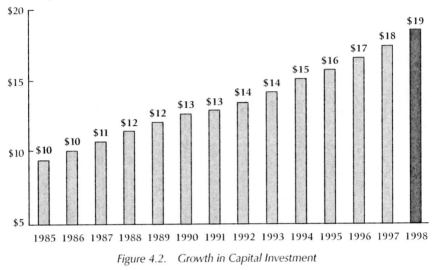

Figure 4.2. Growth in Capital Investment

Percentage Growth in the Number of New Firm Births,
by SIC Code (1989–1996)

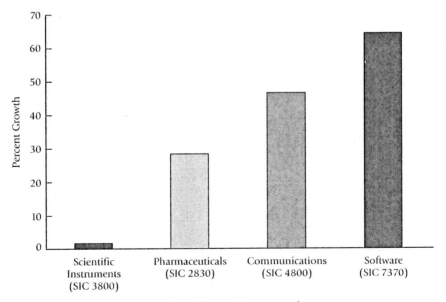

Figure 4.3. Vibrant Entrepreneurship

U.S. GDP Per Capita, Constant 1996 U.S. Dollars

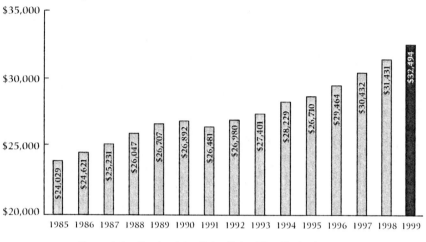

Figure 4.4. Productivity Gains Raised Per Capita Income

While the labor-participation rate in both in-house and two-to-four-year education programs is increasing, the current state of K–12 education does not always provide the necessary base for an increasingly technology-driven workplace (figure 4.5).

It is important to point out that all these changes—that is, growth in technology, increasing capital investment, and the boom of new start-ups with the help of venture capital—have changed the economic landscape. Clusters of new, relocated, or expanded enterprises are forming or have formed in various parts of the country. For example, Washington, D.C. not only houses the federal government and its contractors, but also IT and biomedical companies. San Diego, Atlanta, and North Carolina's Research Triangle Park are other areas that are seeing second and third influxes and growth.

In many cases, these clusters specialize in one or more related product lines and increase their economic power because of the proximity of related enterprises. Clusters are not new creations, but what is new is the increasing number of universities that are at the center of these clusters and are in fact proactive in the formation of new ones: for instance, UCSD (University of California, San Diego) in communication and drug production; the University of Maryland in communication; Johns Hopkins in the life sciences; and Georgia Tech in broadband and wireless communication. Also new are the partnerships forming among competitive enterprises within these clusters, with the universities and government labs being particularly active participants.

Another fact that cannot be overlooked is the dynamism that comes from the interplay of these varied institutions and the movement of people

Erich Bloch

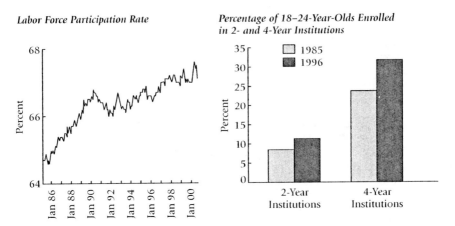

Figure 4.5.   The Drivers of Prosperity: A Larger and Better-Educated Workforce

between them. The Council on Competitiveness, together with Dr. Michael Porter of Harvard, studied these clusters across the United States. The study notes the following observations about regions with clusters in the last decade:

- higher per capita income,
- lower unemployment rates,
- a higher concentration of risk capital, and
- more start-ups.

It would seem that these research parks are an impetus to regional development.

OUTLOOK FOR THE FUTURE

The next question we can ask is, are we as attentive to the future as we should be? After all, as mentioned, the tremendous advances in technology, industry, and economic growth did not occur overnight; they were the result of a formidable investment by the government and private sectors and were driven by geopolitical issues. Overall investment in R & D is a leading indicator of what could occur economically in a decade. We should therefore take a look at this.

Spending for R & D in the United States has been steadily increasing. Between 1995 and 1999, the average yearly growth rate was 6.1 percent, which was about triple the inflation rate. In 1999, the rate of increase exceeded 8 percent from 1998, with funding reaching a level of $247 billion (figure 4.6).

On the negative side, federal R & D has been decreasing as a percentage of the total. In FY2002, federal R & D was about 29 percent of the total; in 1980, it was 50 percent. In FY2000, the federal portion of R & D was as low as 25 percent. The FY2005 budget suggests further cuts are in store. The consequences of this shift in balance between private and public R & D is that the share of basic research, which is the historical province of federal funding, will decrease as a percentage of total R & D being carried out.

The fact that industry focuses primarily on development, rather than research, increases the university's and the government's responsibility to shoulder more of the funding and execution for the basic research component of the R & D equation. This new relationship between the funding sectors also seems less stable. Should there be an economic downturn, industry funding for R & D will decrease, and the country's R & D budget will be more affected than it was in the past, when the two major sources of R & D, industry and government, were split more equitably.

On the positive side, R & D, as a percentage of GDP, is at 2.8 percent (up from 2.67 percent in 2004), which was the highest level ever, except for in the early 1960s. This compares favorably to the R & D spending of our trading partners.

Another observation: throughout the 1980s, federal civilian R & D lagged significantly behind that of the G8 countries (figure 4.7).

If one looks closer into the R & D budget, there are some ominous signals emerging. For example, the allocation of funding is uneven between the various disciplines (figure 4.8). The life sciences are being funded at a much higher rate than the physical sciences and engineering. In fact, the fields with declining dollar amounts of financial support exceed the fields with growing support, fifteen to eleven.

Among the fields that have seen major decreases are physics (20 percent), electrical engineering (30 percent), mechanical engineering (40 percent), chemistry (10 percent), and mathematics (20 percent).

This unequal funding for the basic sciences, mathematics, and engineering is ironic given that a critical component of the foundation for the U.S. technology-based economy is our knowledge of the physical sciences and engineering. Our ability to produce the incredible advances of the twentieth century, such as the transistor, laser, microchip, computer, Internet, and the MRI, has rested on our understanding of the physical sciences and engineering. And in fact many of the advances in the biological sciences would not have been possible without these same developments. Therefore, as federal investment in the physical sciences and engineering declines, not only is the future foundation for these disciplines being weakened, but our economic well-being and the further growth in the life of the health sciences are as well.

While most of these trends are positive for research, let us not forget, however, that competition from abroad cannot be overstated. In fact, R & D

Billions of current dollars

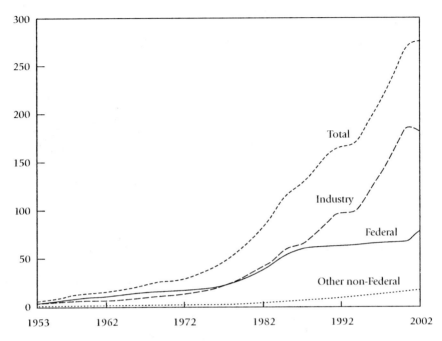

Billions of constant 1996 dollars

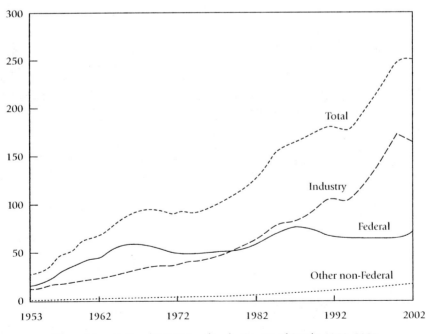

Figure 4.6. National R & D Funding by Source of Funds: 1953–2002

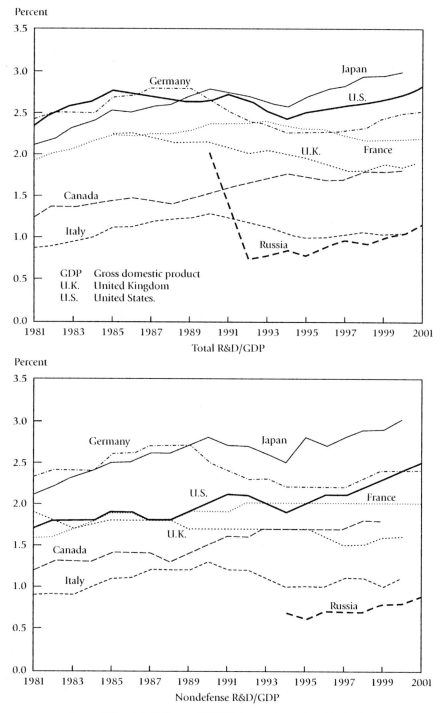

Figure 4.7. *R & D Share of GDP, Selected Countries: 1981–2001*

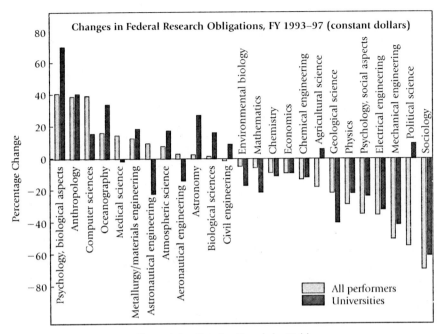

Figure 4.8.   Funding Trends by Field

growth in many countries exceeds that of the United States (figure 4.9). The lineup of countries is also changing. A few of these newcomers to the fore-front are the Scandinavian countries, like Finland, which are increasing their R & D capabilities, as well as some upcoming Asian nations, like South Korea, which are also gaining in their technological and R & D capabilities.

The U.S. industry cannot meet the technological challenges it faces without a continuing supply of highly educated people, and neither can other countries. The availability of an educated workforce is being highlighted by many CEOs as the limiting factor of a country's industrial growth. The shortfall in qualified labor could curtail the expansion of many companies. Here, again, a comparison with other countries, especially our competitors, sheds some light on this situation (figure 4.10). The pool of the U.S. R & D personnel as a percent of the workforce is smaller than that same number in many of the Organization for Economic Cooperation and Development (OECD) member nations. The outlook for the future is not that much brighter, since the pool of scientists and engineers engaged in R & D in the United States is a declining portion of the U.S. labor force.

Putting the whole picture together, one must note that growth in the United States in science and engineering, especially as measured by the number of doctoral and other graduate degrees, has been primarily due to foreign

**Compound Annual Growth Rate in R&D Expenditures
1985–1999**

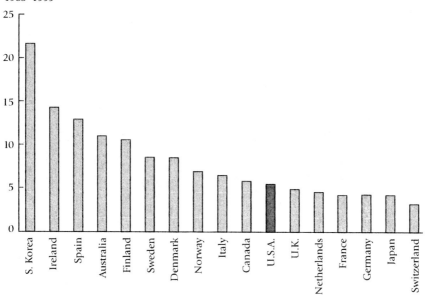

Figure 4.9.   Faster Growth in R & D Abroad

students, not U.S. citizens. Fewer of these international students are coming to the United States in the aftermath of 9/11, and fewer are staying in the United States after completing their degrees, as compared to the past. As a result, they are not joining the U.S. workforce, along with the decline in American students who do not consider science and engineering as their first choices in pursuing a career.

There are many reasons for this state of affairs, and the blame is widespread. Among the more prevalent reasons is certainly the inadequate K–12 education, which does not prepare students for the rigor and prerequisites of a science and engineering college education. Another factor deterring students from studying science and engineering is the underrepresentation of women and minorities in these fields, and consequently the lack of public "role models," although this is slowly changing.

Yet another warning signal comes from comparing U.S. R & D endeavors to those of other nations. Of late, we have not paid much attention to this gauge in light of the sustained—but now diminishing—Japanese and Asian economic malaise. But a number of OECD countries are improving their positions, and the EU, oftentimes seen as an economic bloc, is presenting an altogether different competition than seen from any individual European country in the past.

Total Researchers per Ten Thousand Labor Force

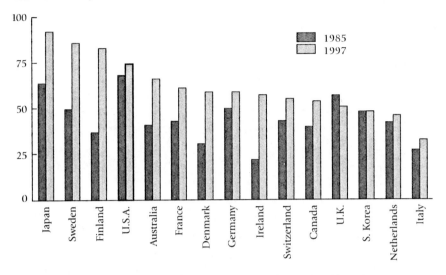

Figure 4.10.   Expanding Commitment to Innovation Abroad

More countries are acquiring the capability of producing state-of-the-art innovations, and new centers of innovation, in fact, are expected to emerge during the next ten years. Many of the countries that were "imitators" in the past are fast becoming innovators—for example, Finland and Korea. This is reflected in the number of high-quality patents (figure 4.11) filed in the United States by other countries. We are seeing advances by countries that, although new to high technology, have acquired an R & D base in their institutions and industry sectors.

Much of this increased effort by the named countries and others is facilitated by the same tools that have contributed and are contributing to our turnaround in prosperity and leadership in technology. A case in point is the use of the Internet (figure 4.12) as it exists today and what it forecasts for the next few years.

Reviewing this comparison of U.S. R & D efforts to those of some of our trading partners, it is worthwhile to keep in mind that we are part of the global economy. One sign of this globalization is the growing proportion that trade represents in the U.S. economy. Both imports and exports are growing at a faster rate than in the past. Since imports exceed exports, the trade deficit in goods and services is greater today than ever. We have looked for excuses: an overpriced dollar, inflated wages, markets that are not open, or oil prices that are offsetting what otherwise would be a surplus. Over time, however, all these purported causes of the trade deficit have been reversed or moderated, and yet the trade deficit continues to increase. The real reason for this trade deficit is

**The Share of a Country's Patents Filed between 1994 and 1998
That Were Highly Cited Patents in 1999, by Sector**

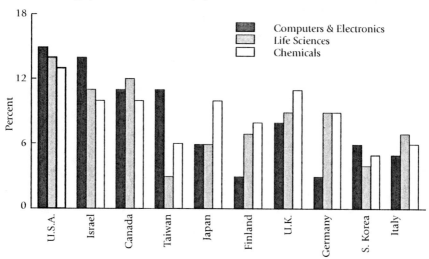

Figure 4.11. High-Quality Patents by Sector

that we are not producing enough high-value products, pointing to a failure in
our innovation system. And yet there is amazingly little concern about this
issue in government or academic circles. This negative trade balance is worth
mentioning, because trade in technology—patents, product licenses, and
copyrights—is positive. In other words, we are providing the rest of the world

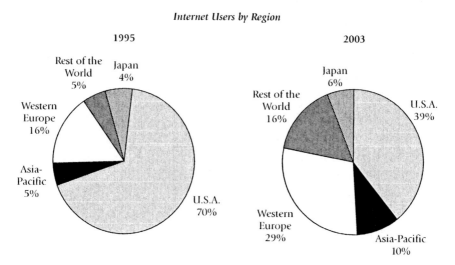

Figure 4.12. Fast Rates of Internet Adoption Overseas

with more intellectual capital than we are receiving in return. While one can legitimately ask if this is good, bad, or immaterial, recent studies suggest that it is precisely this kind of trade that allows other countries to benefit from the fruits of our R & D at an increasing rate.

UNIVERSITIES

The many changes in our innovation system affect universities equally. Universities are an elite group of institutions that have endured for over one thousand years. While true, this statement is also a bit misleading. Today's universities bear little resemblance to those of Paris, Padua, or others that existed in the beginning of the second millennium. And after the dust has settled on this era, universities will be different from what they have been previously, especially the universities in the United States. Both the emphasis and the content of their missions will have changed. Federal and state legislators today, more than ever before, expect universities to contribute to the economic growth of a region or the country. And although this might have been assumed in the past, there is an increasing emphasis on deliverables and results. Peter Likens, president of the University of Arizona, said a few years ago, "If we are to survive we must have sufficient demonstrable value . . . to the people who pay our bills."

On the state level, funding of R & D has increased significantly to nearly $2.5 billion overall. Michigan was an early advocate of this approach, implementing programs to support the state economy more long term. The establishment of a Life Sciences Corridor, with a legislative commitment of $1 billion over twenty years, is a daring attempt to catapult Michigan into the forefront of this field. In addition to furthering science, the legislation clearly keeps the high probability of fallout for economic development in mind. Other states such as Texas, Ohio, and Pennsylvania are considering similar initiatives.

Industry's share of academic R & D increased from 3.9 percent in 1980 to 7 percent in 1998, reflecting the deepening relationship and interdependence between the university and industry. This is a positive development for the universities, since it will enhance their important role in the nation's life.

Nothing is free in life, however. Partnering with industry will put constraints on universities. Universities will have to navigate between their traditional goals of open inquiry with globally shared results and the restrictive intellectual property constraints that company-funded R & D can impose on partnerships. However, the university that can accommodate this relationship will definitely have a competitive advantage.

Other forces are at work in these universities. Academic institutions are moving from an expansionist period to one of restricted growth. The current expansion that the university system is experiencing has been taking place over the course of the last forty years and is an outgrowth of U.S. confrontation and containment policy. In this shielded environment, forces such as unprecedented growth and the increased emphasis on applied versus basic research were set in motion, especially in universities. Today, these same universities live under budget and other constraints.

Secondly, there is an increasing number of institutions laying claim to government funding for research. In 1975, 555 academic institutions were supported by the government; in 1995, 882 such institutions competed for federal funds. We must ask the following questions: Are we investing in the right areas? Is there too much redundancy in the system? Are we promptly terminating research areas and problems that are no longer relevant? What is the uniqueness of a particular institution? These are difficult questions, and each institution needs to answer them for itself.

There is also the obvious fact that multidisciplinary research is playing an increasing role in university research. In the past, the preponderance of university research was devoted to the single investigator/researcher within a discrete discipline, in his own laboratory, all the while working on a problem. While this was the dominant model for university research, centers, subsidized laboratories, and institutes also contributed different research models. However, what is new is the increasing importance of organized centers within the broader university life.

If these centers and institutes are as important as they seem to be, how do they fit into the governance of the university? Departments still determine the curriculum and faculty promotion, and ultimately they still have a veto role over the future of researchers, even when an individual's research is completed within a center.

The increasing use of IT in the university could play an important role in this process of reevaluating the university. The use of technology can potentially bridge both geographic and institutional boundaries. It can provide access to equipment from a researcher's desktop that has not been easily accessible or has required great effort and cost to access. This allows for the possibility of virtual institutions, centers, or groups that could change departmental and institutional relationships, lower barriers, and eliminate duplication in facilities and instrumentation.

It could also give rise to competition. No matter where they are physically located, if it is easy to access instruments and capabilities, the cost-effective provider of quality research will win over the local one who is deficient in service. The same applies to education. Education offerings by nontraditional sources—for instance, for-profit universities such as NTU, an engineering university that offers master's degrees over the air—are only examples of what could happen on a larger scale. What has been said about the impact of research on the economy and the increasingly multidisciplinary nature of research not only applies to the natural sciences and engineering, but also to the arts, humanities, and social sciences.

These areas could potentially contribute richly to cutting-edge research activities, and it would be worthwhile to examine whether propositions allowing for interdisciplinary approaches to solving problems, doing research, and vying for funding would, in fact, even be feasible. Examples of this would include the role of the arts and engineering in media research, increasingly complex legal considerations in systems like networking and communications, and the increasing importance of human factors in design.

The arts and humanities do not always feel welcome in the science and technology environment because we tend to use expressions like "technology transfer" when we really mean "knowledge transfer." The term "technology transfer" excludes the disciplines that we need to attract. This change in frequently used language should stimulate the participation of the humanities, social sciences, and arts in this new age of knowledge transfer, making it clear that their participation is assumed and vital.

CONCLUSION

Universities, above all, along with their faculty, have to realize that they will be expected to contribute to their regions' economic well-being with as much effort as they would expend on education and generation of knowledge. But partnering and collaborating with industry and other institutions is a way to enhance everyone's capabilities.

I would like to conclude with a quote from Charles Darwin, who focused on natural selection and survival of living organisms and said, "It is not the strongest of the species that survive, nor the most intelligent, but the ones most responsive to change." This is an apt and astute observation for all of us today.

Following are Dr. Bloch's responses to questions posed during a discussion of his paper at the University of Illinois Center for Advanced Study.

Q: How does federal government spending on research compare with the money spent by the private sector?

A: *Obviously, in terms of percentages, it is a smaller amount than it was in the past. However, let me give you some numbers: in 2000, the total budget for U.S. R & D was about $240 billion, 28 percent of which came from the federal government, and over 70 percent came from the private sector. That is a completely different relationship from what you would have seen ten to fifteen years earlier. In fact, in 1985, the total R & D budget of the United States was 50 percent from the federal government and 50 percent from the private sector. So a completely different relationship has been established which has both some positives and negatives.*

Q: What do you think are some of the positives as well as negatives of these changing relationships?

A: *A high dependence on industry for R & D funding can be a problem. As we have seen in recent years, the first thing that is going to happen in industry during an economic downturn is cutting R & D, because the savings fall directly to the bottom line, and consequently the cut affects the bottom line directly. This can mean a major reduction in U.S. R & D during down cycles in the economy. The second problem with an overwhelming portion coming from the private sector is most of the funding is spent on development, and very little for basic research. The bottom line, therefore, is that the money allocated for basic research is a diminishing part of the country's R & D budgets. This is a bad sign because the research engine needs to do its part in order to have development and consequently new innovation and products in an increasingly competitive marketplace.*

Q: Is it becoming more difficult to make the argument that we should be spending money on pure science, on pure research, that may not yield anything more than increased knowledge?

A: *Well, let us argue that it was never very easy to make that argument in the first place. Because what you are hinting at is that there is pressure on the universities today to show results, that is, results in terms of economic competitiveness, for example. And yes, that exists. And no doubt that pressure by the states as well as by the federal government and industry will continue. While this is not ideal, it is not all bad either, because what is happening today is an acceleration of development and a commensurate acceleration of new ideas. And when that happens, there is no clear demarcation line between R & D, because the two develop together. Therefore, I think it is more appropriate that one asks the question, what are researchers doing today, or what is the thinking today that can affect or contribute to the economic competitiveness of the country or region?*

Q: In what way(s) are the priorities of universities changing as they think about what their role is in research?

A: *I think both the roles and priorities of the university are changing in many different ways. First of all, universities today are not only concerned with individual investigative research; they are very much concerned with multidisciplinary team research. This should involve the total university, not just science, but engineering and science combined, along with the arts and humanities. The stovepiping of yesteryear, so to speak, needs to be pushed back into the background, and the university as a whole has to essentially work closer together. The university as an ivory tower is no longer acceptable. The university has to partner with industry, the government, government labs, not-for-profit organizations, and institutions.*

Q: Within the last few years, there has been some controversy/discussion within the scientific community about what it is we're spending our money on—in particular, questions about some of the very large projects. The most high-profile of such projects is the human genome mapping effort. There was also one point when we were talking about building an expensive supercollider. Eventually, Congress got to the point where they said, "No, we are not going to do that." But I think a lot of people are concerned that when you have a few large projects, like the genome project, taking up a lot of money, it means that money is being taken away from smaller projects, things that might be worthy of funding but are not quite as high profile. Is this still an issue in people's minds?

A: *This has always been and always will be an issue; no doubt about it. But I do not think you can look at it in those terms. Oftentimes, the terms or the premises are misstated. There are projects that require large amounts of funding in order to accomplish them. A telescope costs a lot of money; it costs more than the small microscope that you are used to. But you need a telescope if you want to do astronomy, and there is just no other way of doing it. The question really is, do we have the right balance between large and small science?*

Are we funding the right number of large programs and the right number of small programs, and do all the dollars add up? And it is a juggling act between those two. These are forces that exist within the science and engineering funding agencies, and I think they are the ones that are being given a lot of thought by the university community itself, as well as by the agencies, and in the end, there has to be a compromise. You cannot live with one or the other, because we need both.

Q: I read a comment that was made by Congressman Ehlers (Michigan) who is a physicist and was giving a talk to some of his fellow physicists at Fermi Lab last year. He said something to the effect that he thought that his fellow scientists had not done a terribly good job of speaking for their research and its benefits and the value of promoting it and funding it. And one wonders whether scientists need to be a little bit better at public relations, in talking about what it is that they do and why it is important?

A: *He has made that point many times and rightly so. I agree that scientists and engineers are oftentimes reluctant or incapable to tell their stories, but I think it is crucial that they do. How else does one attract people to the sciences, and second, because funding going to science has to be justified, and there is no better individual who can justify it than the one who does the science.*

NOTES

1. U.S. Department of Commerce, "U.S. Worker Productivity by Industry (1990–1997)."

5

Back to the Future: The Increasing Importance of the States in Setting the Research Agenda

Larry Smarr

My particular approach to the topic at hand is that of an empiricist, presenting data that I have come by over the last fifteen to twenty years while trying to create new institutional frameworks. My aim is to provide glimpses into several working examples of university partnerships with the private sector, state and federal governments, and private donors.

The initiatives I will discuss in the second half of this chapter probably mark the highest level of state initiatives defining the country's research agenda. Whether or not these initiatives can be duplicated remains to be seen. Concerning the programs in planning in California, they are high-risk experiments on the part of California's governor, legislature, and university system. But I am interested mainly in conveying the differences between this approach and the large-scale federal approaches we have come to use since World War II. My title, "Back to the Future," refers to the fact that land-grant universities were founded, not within the last 50 years, but more accurately in the last 150 years. The University of Illinois, a land-grant institution, was born in 1867, along with the University of Missouri, during a time of great political and national turmoil.

When we look at the context in which these universities were established, one question that comes to mind is, why did the citizenry, just after the most deadly war in our history, think it was appropriate to set up these universities? However, it was during this time that the three basic elements we now view as the pillars of the university were formed. First, there was a strong desire to educate a larger workforce. Land-grant universities were then set up as an antielitist educational forum. They went to great lengths explaining how they

wanted to make education available to all citizens of the state. They understood that the workforce had to be educated in order to harness the twin technological revolutions taking place in the productive fields of agriculture and
manufacturing. They wanted to make sure that these land-grant universities
focused on agriculture and manufacturing, and more importantly, on
broadly educating the citizenry.

Why did universities perform research then? Because there was an industrial revolution under way that was completely changing the face of agriculture and making the United States a manufacturing world power. The
Morrow plots[1] are symbolic of how research in the land-grant university
increased the wealth-creation capability of the state's economic sector. Back
then, the extension service was really the birth of service. Researchers did not
stay in their "ivory towers"; that simply was not the nature of land-grant universities. The ivory towers were thought to be places where people only studied the classics and were far removed from the everyday man; in fact, they
were distinctly set apart from these land-grant universities, which found their
niche among the masses.

Today that tradition continues, and one of the core values of the research
university is outreach. Research is associated with wealth creators, new technologies, students' education, and capabilities. The Morrill Act, itself, offers
an interesting parallel when one takes a close look at it.[2] It explicitly talks
about federal funds being used to help the state universities. We think about
federal funds being channeled to help state universities as a post–World War
II phenomena, particularly post-Sputnik, but it was actually established a lot
earlier.

The motto of the University of Illinois, "learning and labor," reflects this integration of the university and the private sector. The plow in the university's seal
was the top-notch technology for those in manufacturing at that time. If we
built a university today, most likely the computer and the Internet would be on
its seal. This is not something divorced from the university; that is, there is no
great wall blocking the journey from the university to the economy. Rather, the
university and the economy exist together in the seal as part of the university.

Figure 5.1 is a timeline from 1965 up to 1994 from the National Research
Council's Brooks/Sutherland Report that I was privileged to work on.[3]

The black lines show when the billion-dollar-per-year industry came into
being. The NSF and Defense Advanced Research Projects Agency (DARPA)[4]
support researchers with time sharing, graphics, networking, workstations,
Windows, risk microprocessors, very large-scale integration design, and ray disk
parallel computing. What you can see here is that the federal government
funded research in universities ten to fifteen years before those ideas led to $1
billion plus in new industry. In the post-Sputnik era, the federal government's
funding of research in universities led to what we now refer to as the IT and
telecommunications industry. It certainly is the case that monopolies like

TOP

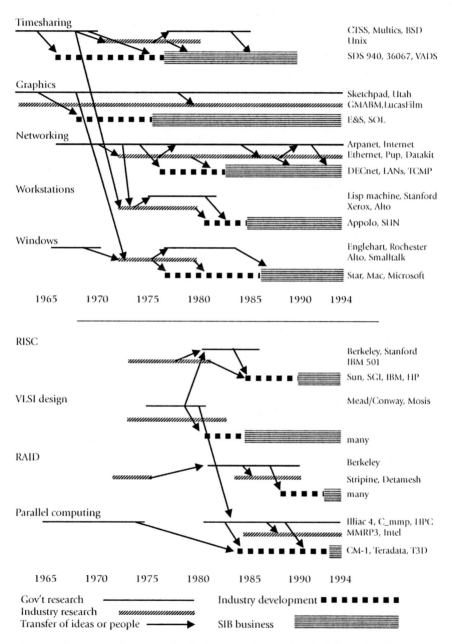

Figure 5.1.   The Critical Role of Federal Funding in Creating the Information Economy

AT&T, or virtual monopolies like IBM, developed great research labs during this time. Ironically, the resulting technology, particularly the Internet and e-mail, led to the demise of many of these industrial research labs during my time as director at the National Center for Supercomputing Applications (NCSA).

Thus, the question arises, what should the federal government's role be in the IT and telecommunications research arena? The President's Information Technology Advisory Committee (PITAC)[5] was asked to consider this question essentially de novo. The committee was put together in 1998 and was the highest-level group in the country looking at the issue. Half of this thirty-person committee was from industry—CEOs, CTOs, and vice presidents—and the other half was from academia. For symbolic purposes, the first cochairs of the committee were Bill Joy, the cofounder of Sun Microsystems, and Ken Kennedy, an outstanding computer scientist working with parallel computing from the university sector.

We saw the destruction of these central industrial labs. We had a great deal of information. We took in many hours of testimony and depositions by all the participating agencies. In the end, a particularly important portion of our report was a two-page sidebar, which was written by the industrial members of the committee and is now available on the Web.[6] It was an industrial manifesto completely affirming the notion that the federal government's research priorities lie in long-term, basic research funding for universities in information technology.

The conclusion is then that corporations could not do it. As the private sector, they are not capable of doing long-term research because of the nature of the competitive forces in today's global economy. A dollar that does not go toward increasing the next quarter's profit and creating shareholder value is misspent because stocks will go down. This pressure has transformed the private sector to the point where they are only capable of short-term research. And, as for industry, 90 percent of the R & D budget is the development component, and any research that takes place is short term.

I believe that we live in a post-PITAC era in which industry understands the value of universities and of the federal government's role there. When PITAC was first started, people in Congress said that IT was the last place that the federal government needed to spend money because the private sector had no lack of funds. If industry wanted more ideas and increased innovation and discoveries, then they should "foot the bill." After the PITAC report, however, those arguments were put to rest. It took a lot of testimony and other evidence, but Congress became convinced of their primary role in the funding of long-term, basic research in universities. That was an important milestone.

In addition to supporting basic individual principal investigator (PI) research, the federal government has funded large-scale activities, such as the National Computational Science Alliance (the Alliance) or the National Partnership for Advanced Computational Infrastructure program (NPACI), which join many universities via partnerships, grants, and computational infrastructure.[7] Fifteen or twenty years ago, the picture was vastly different. There was no access to supercomputers in this country. The Internet was not widely available. The NSF-net was built on top of the Advanced Research Projects Agency Network DARPA-net, which was Department of Defense technology.[8] The NSF-net first connected the five supercomputer centers, then later the universities, and then ultimately became the global Internet. There was no scientific visualization,[9] and the Mosaic Internet Explorer was just a patchy server used at NCSA.[10] All of this is the information infrastructure that the world and the economy has come to depend on, and it was the result of the federal funding provided to universities.

If I look back on my fifteen years with the Supercomputer Program and PACI, I think about the partners.[11] Out of the twenty thousand grants that the NSF has, PACI is probably the largest single PI grant, so it is worth examining. Here, the federal government clearly has the primary funding role, and as a result, it defines the program, designs the competition, chooses the groups to choose the winners, and evaluates everything. The state really has no role in this process. On the other hand, NCSA and the Alliance could never have accomplished what they did without the core cost sharing that came, and continues to come, from the state. The state was the major source of funding for the core salaries at NCSA, the leading-edge site of the Alliance.

Industrial partners were also incredibly important. Approximately $60 million over the course of fifteen years came from industrial partners to the NCSA. These industrial partners are users of IT and include Caterpillar, Sears, Allstate, Boeing, American Airlines, and Eli Lilly. These are all of the companies that make up the economy that consume IT and telecommunications. These were the companies that were attracted to this project. (This is in stark contrast to the California Institute, where it is not industrial partners, but rather creators of the IT and telecommunications industry, that are partnering with us.)

The University of Illinois at Urbana-Champaign (UIUC) has a very interesting set of values that allowed these companies to do secret proprietary research on campus, with full levels of industrial security. Many campuses do not allow this. With justifiable concern about faculty and graduate students being able to publish, UIUC has figured out a way to make this possible. As a result, many of the things that the NCSA is famous for came about because of initiatives and projects no one will ever know about. New software capabilities were required to solve difficult problems that multibillion-dollar corporations could not solve inside their companies. When the program was started, many in industry were horrified at the idea that something so secure was in such an insecure environment. Yet the benefit must have outweighed the cost, because they risked bringing their crown jewels to this campus. This campus now has become an engine for development.

The NCSA is not only beneficial to the university and industry; we were also able to reach out into the entire Champaign County community through CCNET.[12] We worked with the community on a scale never seen before. At least two hundred community leaders from both Champaign and Urbana (C-U), and within Champaign County, volunteered to be a part of CCNET. It made C-U one of the earliest electronic communities. Through this initiative, computers were put in the libraries for the first time, ISDN was installed in all twenty-seven schools in both school districts, competing local medical complexes adopted common IT standards, and small businesses also started using the system. This was a successful, completely volunteer effort between the campus and the community. And it was mediated and brought into being by the existence of this large-scale, federally funded structure, with strong cost sharing from the state.

Having said that, I have to come back and point out that from the federal government's point of view, there was no mandate that this large entity in the center of the state could be used to enhance the competitiveness either of this community or the state. In fact, one might even argue that such a mandate was excluded. The NCSA existed to serve the entire country. There was a sense that one should not have preferential treatment locally. I think this had a lot to do with why one could argue that the state of Illinois did not gain a corresponding advantage competitively even though there were only two of these research centers in the entire country: one here and the other in California. It had a lot to do with the partnership structure, which is really what this chapter is about: how structure develops partnerships, and the consequences of these partnerships.

Another paradigmatic example from UIUC is its interdisciplinary Beckman Institute.[13]

Here, it is neither the federal government nor the state that is the primary funder, but private individuals, the Beckmans, who provided the resources that made the institute possible. The state again provided cost sharing in recurring funds. The money was used for buildings and world-class facilities.[14] After some time, there was a realization that Beckman needed to go from twenty-one different programs to three superfocus areas. Conceptually, it helped explain why people were actually at the Beckman Institute, while increasing the chances of getting federal funding. The institute could bring larger-scale teams together to address more complex problems, and it therefore became a framework for taking existing faculty and fusing them together into teams in a way not possible with departmental and college structures.

Former governor Gray Davis (CA) launched a similar type of initiative that is quite large in vision, scope, and funding. It may be the largest such initiative any state has ever undertaken, although funding has declined since the more recent onset of California's economic woes. Davis reasoned that Silicon Valley had become the engine for California, driving it to the point that California now has the sixth or seventh largest economy of any country in the world. And he noticed that Stanford and Berkeley seem to have had a great deal to do with creating this. For example, Mr. Hewlett and Mr. Packard were given land by Stanford that eventually became part of Silicon Valley. Moreover, many ideas, which eventually developed into companies, were first conceived at these universities; for example, SUN is an acronym for the Stanford University Network that was prototyped by Andy Bechtolsheim and others, Jim Clark was an assistant professor at Stanford before forming Silicon Graphics, and so forth.

I think Governor Davis borrowed an idea from the federal paradigm when he decided to foster competition among the nine campuses. The stipulations were as follows: they could choose any topic they wanted that would improve the future competitiveness of California, either a research area or something that would improve the quality of life of the people of California, and they could team up between two or three universities to do it. Eleven proposals went in, and six were selected for full proposals. Three were then fully funded during the 2000–2001 school year, while one more proposal, led by Berkeley, was still pending. And the topics spanned what most people say will drive the twenty-first century: nanotechnology, biomedics, IT, and telecommunications. In fact, there is great overlap among the edges of all these areas.

In the case of the establishment of the physical infrastructure, the state sets the agenda, provides the fundamental funding, and sets the competition. These funds, as in the private-donor case of the Beckman Institute, are appropriated for new buildings and equipment. There was a proposed operational budget comparable to what the State of Illinois provides to NCSA, but as a rule, the governor had a four-year, two-to-one cost-matching plan. That is, for every one of his dollars, the proposed idea would need to bring in two more dollars over the course of four years. There was a sense that because of the initial funds, the new facilities, and the teaming together of individual faculty, you would create something powerful, increasing the competitiveness for federal grants and generating a lot of the cost sharing from industry.

The resulting center we put together spans the University of California at San Diego and Irvine. This is along what is called the Southern High-Tech Coast. The northern part of High-Tech Coast goes up past UCLA, all the way to Santa Barbara. That is where California's great growth is predicted to be in the future. Within this funding model, the governor allocated $100 million per institute for capital. This money is funding construction on both campuses and an industry investment of $140 million over the course of four years.

Our focus on this project is on the set of technologies that go beyond today's wired PC endpoint Internet. We will study what happens when the market for broadband services goes from a few million homes to tens of millions of homes in the next few years. There are currently half a billion PCs in the world, but we are going to have billions of new endpoints, sensors, actuators, handheld devices, and so forth. All of these new endpoints will drive data going into the Internet, so the core of the Internet will have to go into an all-optical network.

Furthermore, because of broadband connections to many home PCs, we expect to see the emergence of a planetary computer. There are many companies that are working with us, as well as others across the country, including Condor, coming out of Knowledge Port here, who are working in that space.[15] Effectively, the end result is a mobile Internet that is available anywhere, anytime, and is powered by a planetary-scale computer for computing and storage. We worked this out by essentially calling for faculty that would be interested in this endeavor. Basically, we are working with faculty from the University of California at San Diego (UCSD) and the University of California at Irvine (UCI) to create teams that can take a whole system engineering approach to a world of technology that does not yet exist. As I said, this is an experiment, and one that has a high probability of failure.

About 3 to 4 percent of the faculty at UCI and UCSD, about 220 individuals (table 5.1), signed up for this adventure and organized themselves into multidisciplinary teams. By taking a cognitive trip into the future, we were able to come up with the subcategories of researchers that we had to have. These new technologies will not require Pentium IVs and little button-size sensors but a whole new class of materials and devices. We are going to build the living laboratory of the future and its networked infrastructure. Because of all the software and human-computer interface issues, about half of the faculty is from electrical engineering and computer science. We are taking this new, standards-based Internet and customizing it into four larger areas of research, which happens to span about 80 percent of California's economy: environmental and civil infrastructure, intelligent transportation systems of the future, digitally enabled genomic medicine, and the emerging new media arts.

Moreover, it is important to consider that, from the start, our planning included people in policy management to prepare for the socioeconomic evolution that this revamped Internet is going to cause. Industry and education could become interwoven at all levels.

Table 5.1. System Integrated Approach Focusing on Intersections

EDUCATION	POLICY, MANAGEMENT, & SOCIOECONOMIC EVOLUTION				INDUSTRY
	Environment & Civil Infrastructure	Intelligent Transportation	Digitally Enabled Genomic Medicine	New Media Arts	
	INTERFACES & SOFTWARE SYSTEMS				
	NETWORKED INFRASTRUCTURE				
	MATERIALS & DEVICES				

Table 5.2. Large Partners > $10M over Four Years

Akamai		Mission Ventures
Boeing		NCR
Broadcom		Newport Corporation
AMCC		Orincon
CAIMIS	**Computers**	Panoram Technologies
Compaq	**Communications**	Printronix
Conexant	**Software**	QUALCOMM
Copper Mountain	**Sensors**	Quantum
Emulex	**Biomedical**	R.W. Johnson Pharmaceutical RI
Enterprise Partners VC	**Startups**	SAIC
Entropia	**Venture Firms**	SciFrame
Ericsson		Seagate Storage
Global Photon		Silicon Wave
IBM		Sony
IdeaEdge Ventures		STMicroelectronics
Intersil		Sun Microsystems
Irvine Sensors		TeraBurst Networks
Leap Wireless		Texas Instruments
Litton Industries		UCSD Healthcare
MedExpert		The Unwired Fund
Merck		WebEx
Microsoft		

One of the big differences between putting the California Institute together between San Diego and Orange Counties and putting the NCSA and the Alliance together in Champaign County is the scores of telecom companies in San Diego and Orange County. These companies form the San Diego Telecom Council, whose aim is to brand San Diego as the telecom center of the universe (table 5.2).

Many industries will develop at the intersection of the fiber. Underneath downtown San Diego, there are seventy thousand strand miles of fiber—this is ten times the diameter, or three times the circumference, of the earth. Other industries emerged this way. Chicago came to be at the intersection of rail lines. St. Louis came into existence at the intersection of rivers. We are using the San Diego Telecom Council and UCSD Connect, a wonderful tech transfer program that has been developed at UCSD, as funnels into this larger community. There are other programs and centers as well: Biocom transfers technology to biotechnology companies, there is the UCI Chief Executive Roundtable,[16] and there are dozens more. The point is that there already existed a well-organized commercial environment into which we could build.

This is part of how I think we can enhance technology transfer. There are a lot of venture capital and other investment and incubator activities already going on in the San Diego area. We approached them, asking them to partner with us for pro bono services. There are also many pre-IPO companies who have new widgets that would fit into this new Internet but have no way to test it at scale, which partnered with us by giving us early access to their technology.

We will also form an investment council between university intellectual property officers and tech transfer people like UCSD Connect. In fact, this council will probably be the facilitators of this investment council and the larger-scale investment community. When a faculty member has an idea, we will know a priori which partners to call so that the technology discussion about the transfer of that technology to the private sector can be shortened, as well as making many more channels available and open to the faculty. This will also make it much clearer to the investment community which new ideas are emerging out of the university.

Notice that the partners are not the users of the technology the way they were within the federal program structure. Most of the industrial partners in NCSA's program were chosen a priori because they were among the top companies in the nation within that sector of the national economy. Here, we were mainly focusing on the California economy since it is a state-led initiative. We had that mandate. We were able to cover the various components, which was seen earlier in the faculty research, with similar components of computer software, telecommunications, and so forth, in this private sector. One of the goals, then, was to weave together the interested parties on both sides.

Another fascinating thing about this experience is seeing how industry has taken a more long-term perspective on why investing in universities leads to return. Look at the allocation of the $140 million that was invested.[17] The money has partially been allotted for endowed professorships, graduate-student fellowships, equipment donations, named laboratories, and pro bono services, all of which are more long-term, lifetime return investments (figure 5.2). Sponsored research is also there, but only one-fourth to one-third of the resources that traditionally go into sponsored research has been allocated as such. And of course that sponsored research is under the nominal review of the university the way any other such sponsored research would be. This is not groundbreaking work, but there is no preferential treatment for these companies.

I would like to provide some examples of the early work that the California Institute is currently engaged in. First, it is built on the existing strength at the various universities. For instance, there are twenty faculty members doing wireless communication research with the Center for Wireless Communication. That is a part of the institute. At the Irvine campus, there is a BioMEMS facility (MEMS stands for Micro/Micron scale Electro-Mechanical Systems). But micron-scale and millimeter-micron-scale devices, such as DNA arrays, are very large compared to where we are going. If you magnify that DNA array by five hundred times, you get down to some of the work Karl Hess has done

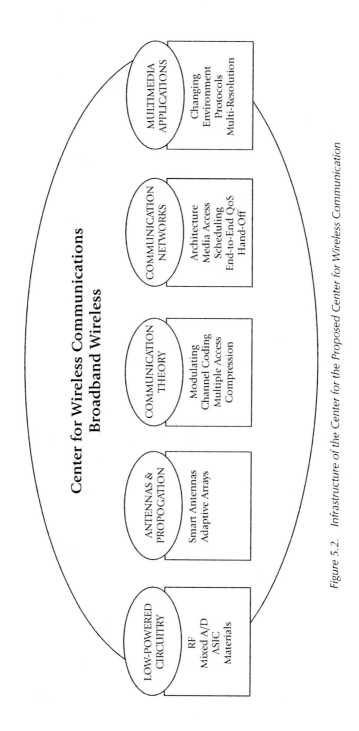

Figure 5.2.   Infrastructure of the Center for the Proposed Center for Wireless Communication

at the Beckman Institute. We are also going to build clean-room facilities in our institute going down to nanoscales, which are four hundred times smaller than a DNA array. These solid-state lasers actually require quantum mechanics to function.

The rhinovirus, which causes the common cold in humans, is considerably smaller than five nanometers, as is an IBM quantum corral, the quantum wave functions of individual atoms sitting on copper.[18] At the nanoscale level, atoms are about a tenth of a nanometer. Both of these devices, one "organic" and the other "inorganic," are capable of storing bits.[19] In the virus, it's called RNA or DNA, and in the IBM constructs you can capture things in the corral. At this scale, the disciplinary boundaries that we have spent so many years developing fade away. Here, there is no biology, physics, or engineering, just a pile of atoms. I believe a great deal of the invention in the twenty-first century is going to occur at this scale. A third of the new building costs at UCSD is going to fund clean-room facilities allowing this type of invention to move forward.

We are also working with the Center for Research in Computing and the Arts (CRCA) to build performance spaces.[20] CRCA is a unit at UCSD that searches out many of the ways in which information technology alters the arts. We are working with them and similar programs on the Irvine campus, like the Bell Center,[21] together with those working with experimental music technology, theater, and dance departments, in efforts to weave together one of our four great application challenges—that is, to understand how art forms, and in particular computer games and cyberspace civilizations, are being created by people to live their lives preferentially compared to their "less acceptable" lives in physical space. Even the La Jolla Playhouse is interested in doing this, because CRCA is cutting edge. After training new students in these departments and then in community playhouses, these art forms have become a part of plays, many of which have gone on to become great Broadway hits. This avenue is one way to get into both the economy and the culture.

We are also building a sixth undergraduate college solely devoted to arts technology and sciences (figure 5.3).[22] This will be a part of our living laboratory. We are planning to create this living lab of fiber optics for the new wireless Internet at this undergraduate college.[23] People are just now getting gigabit Ethernet, connecting at a billion bits per second. We want to start with eight gigabits, move on to eighty, and then go to one thousand gigabits in fiber over the next few years. Of course, this could not be done without our industrial partners.

We are going to be blanketing the entire campus with broadband wireless. In fact, Irwin Jacobs, the founder of Qualcomm, gave $15 million as the first gift to start the California Institute, saying that the most valuable return on his investment, in terms of creating future wealth for his company, was to

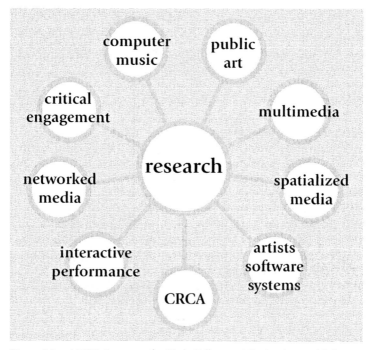

Figure 5.3.   UCSD Becomes the Sixth College to Mix the Arts and Sciences

provide funding allowing twenty thousand young people to figure out how to use this new Internet and then sitting back and watching them.

Similarly, the San Diego Supercomputer Center (SDSC) has also been doing work with wireless broadband Internet in collaboration with the Alliance—the so-called 802 that is all over the campus and putting eleven megabits per second in the air.[24] Prior to the California Institute, Hans Werner Braun at the SDSC built a forty-five-megabit-per-second device. The SDSC and the Scripps Institute of Oceanography got a $2 million grant from the NSF for work in the Mission Bay area.[25]

Palomar Mountain currently has access to forty-five-megabit-per-second point-to-point wireless. The Scripps Institute of Oceanography is now linking all of their sensors together. All this requires is for someone to climb up these areas and put antennas up. This wireless infrastructure goes across all kinds of areas where there is currently no communication infrastructure.[26]

There are twenty-five thousand bridges in California, none of which are instrumented.[27] Every time there is an earthquake, the National Guard comes in, and the problem is that they have no idea which streets, bridges, and buildings are either stable, beyond hope, or have high needs. This is not because we do not have the technology. If these structures are instrumented and all the time measuring the intensity every time the earth shakes, then all smaller-scale occurrences are simply catastrophe prevention, or at the very least, catastrophe anticipation. Neural nets, for instance, will work on the data that is generated once per second from these areas or structures to define whether a particular bridge is unstable or not. One pillar might be on quicksand and the other one on solid rock. We would get second-by-second readouts of the life history of the stimulus of the earth motion, big trucks going across, and ensuing responses.

The point is that when an event that is out of the ordinary occurs somewhere, all of these data sets are being analyzed away from the event. We know instantly which ones crossed a threshold exceeding their carrying capacity and probably broke, and which ones just got a good shaking but are otherwise perfectly fine. We are installing these instruments in Southern California. We do not have to do the whole country because we have a mandate from the state to focus on it, and primarily Southern California. This makes a lot of things possible that are not the case if you have to look at the entire country. We are building large-scale facilities for analysis with our industrial partners. We will be optically networked between San Diego State University, UCSD, UCI, and soon back in Illinois via the PACI-net.

We do not want to reinvent anything that the NCSA, SDSC, the Alliance, or NPACI already have done. These things will be adopted and then worked out further in the future with the two hundred plus faculty researching different content areas. We are going to work with the community on whether an instrumented environment and an intelligent transportation system can avoid some of the negative side effects of growth.

For example, about a million people are predicted to come to San Diego County over the next ten to twenty years. That amounts to approximately 700,000 additional automobiles and the creation of 1,000 miles of new parking lots in order to provide one parking space for each of these new cars, not to mention all of the traffic. We must find ways to use intelligent transportation to limit the traffic jams. We have to ask ourselves the what-ifs to figure out where to build the highways. This also requires knowing the water cycles, the ecological systems that might be disrupted, and so forth. This type of work and planning requires a major long-term involvement with the community.

Finally, of course, we are competing for big federal grants—that is, National Institutes of Health (NIH) grants. I have been the adviser to the director of NIH from the IT perspective for the last five years and have coauthored a report on the needs of the biomedical research community. One of the first responses of the NIH was to create a deep web of information at the leading brain-imaging research facilities in the country, from Harvard to Caltech and Stanford. The goal, then, was to take and build a software infrastructure across the network connecting these machines that would federate the brain-imaging data to increase knowledge about the brain at multiple scales.

Each brain is unique to the individual, and currently, if you were to go in and get scanned at one of these places, your data would stay at that place, if it would be stored at all.

For the first time, this project would create a national library of brains, with all of the variations that occur, and the ability for the first time to study that variation. At the same time, the genome project is bringing all the new data about the differences in individual genomes to light. Again we can see overlaps between the California Institute, SDSC, and the Center for Research in Biological Structure (CRBS), funded by NIH and run by Mark Ellisman.

In sum, the IT field continues to look forward to the future between researchers and the private sector. The difference between the NCSA, the Alliance, and the California Institute is not its subject matter, but its infrastructure, funding, and what the mandate is. And in the end, the partners are the same: the federal government, the states, the universities, and the private sector. The only difference lies in who takes the initiative, who partners it, and for what those partnering assets are used. It will be these differences, then, that will lead to the development of new models.

NOTES

1. The Morrow Plots, first cultivated in 1876, are the oldest agronomic experiment fields in the United States. They are located near the center of the University of Illinois Urbana-Champaign.

2. The Morrill Act of 1862, 7 U.S.C. 301 et seq., established the land-grant university system. The legislation introduced by U.S. Representative Justin Smith Morrill of Vermont granted 30,000 acres of public land for each senator and representative under apportionment based on the 1860 census to each state. Proceeds from the sale of these lands were to be invested in a perpetual endowment fund that would provide support for colleges of agriculture and mechanical arts in each of the states.

3. See slide 4 of 32 at www.jacobsschool.ucsd.edu/~lsmarr/talks/UIUC%20CAS .4.01_files/frame.htm. See also http://bob.nap.edu/html/hpcc for the full Brooks/ Sutherland Report.

4. DARPA is the Department of Defense's central research and development organization.

5. See www.hpcc.gov/ac/about.html.

6. See www.ccic.gov/pubs/pitac/index.html.

7. See www.npaci.edu. NPACI is a large-scale computational infrastructure building partnership similar to the Alliance. It is led by the San Diego Supercomputer Center (SDSC).

8. NSF-net was a federally provided, general-purpose backbone network for the research and science community that closed in 1995. Its roots stem from early ARPA research on packet switching and development of the TCP/IP protocol suite.

9. In fact, Stefen Fangmeier, who was the head of our Scientific Visualization Group in 1987, is now the premier wizard at Industrial Light and Magic, and was

given the honor of probably being the first computer graphics person listed in open-ing credits of the film, "The Perfect Storm," along with George Clooney and the other stars.

10. Mosaic was the first readily available graphical Web browser, released in 1992.

11. See www.ncsa.uiuc.edu/About/PACI. Partnerships for Advanced Computa-tional Infrastructure (PACI) is a National Science Foundation program started in 1997 to create the foundation for meeting the expanding need for high-end compu-tation and information technologies required by the U.S. academic research commu-nity by developing the new national computational infrastructure called the Grid. It unifies the Alliance and NPACI under one program.

12. See http://archive.ncsa.uiuc.edu/Edu/trg/ccnet.

13. See www.beckman.uiuc.edu. The Beckman Institute for Advanced Science and Technology at the University of Illinois Urbana-Champaign was established with funds from the Arnold and Mabel Beckman Foundation. It is an interdisciplinary sci-entific research center.

14. See www.beckman.uiuc.edu.

15. Condor is a Knowledge Port product that coordinates the available power of networked clusters of workstations and PCs.

16. See Esther Schrader, "Meet Orange County's New Power Elite," *L.A. Times*, October 26, 1998, A1. The UCI Chief Executive Roundtable was established in 1986 to encourage cooperation between business and education at University of Califor-nia Irvine.

17. See slide 18 of 32 at www.jacobsschool.ucsd.edu/~lsmarr/talks/UIUC%20CAS .4.01_files/frame.htm.

18. See slide 20 of 32 at www.jacobsschool.ucsd.edu/~lsmarr/talks/UIUC%20CAS .4.01_files/frame.htm.

19. See slide 20 of 32 at www.jacobsschool.ucsd.edu/~lsmarr/talks/UIUC%20CAS .4.01_files/frame.htm.

20. See www-crca.ucsd.edu.

21. See http://bellcenter.uci.edu/hm.htm.

22. See slide 22 of 32 at www.jacobsschool.ucsd.edu/~lsmarr/talks/UIUC%20CAS .4.01_files/frame.htm.

23. See slide 23 of 32 at www.jacobsschool.ucsd.edu/~lsmarr/talks/UIUC%20CAS .4.01_files/frame.htm.

24. The IEEE 802.11 is a standard wireless transmission protocol defined by the Institute of Electrical and Electronic Engineers.

25. See slide 25 of 32 at www.jacobsschool.ucsd.edu/~lsmarr/talks/UIUC%20CAS .4.01_files/frame.htm.

26. See slide 26 of 32 at www.jacobsschool.ucsd.edu/~lsmarr/talks/UIUC%20CAS .4.01_files/frame.htm.

27. See slide 28 of 32 at www.jacobsschool.ucsd.edu/~lsmarr/talks/UIUC%20CAS .4.01_files/frame.htm.

6

Global Public Goods for Poor Farmers: Myth or Reality?

Timothy G. Reeves and Kelly A. Cassaday

At the start of a new century, the international agricultural research and development (R & D) community is undergoing a transformation. Powerful forces are acting to expand research opportunities as never before, but at the same time they seem to have raised barriers to research that are greater than any that have been seen in the past. For many years, international agricultural research organizations have worked very effectively to improve the lives of poor people in developing countries. As research funding diminishes, and as quiet scientific controversies become incendiary public debates over patenting life forms and rights to genetic resources, many are questioning how much longer international agricultural research can continue to help poor people. International agricultural research has provided improved seed, better agricultural practices, and information that have helped poor people immeasurably, but the rules of research are changing. Will the new rules transform these so-called "global public goods" into vanishing commodities, or into commodities that poor people cannot hope to access? That is the central question explored in this chapter.

The vast majority of the world's poorest farmers still produce crops using farm-saved seed and traditional crop management practices that have been passed down from generation to generation. These can be regarded as a form of "global public goods." Before we discuss why global public goods are important for the world's poor people and whether developing countries will have access to them ten or twenty years from now, it is useful to explain what we mean by "public goods" and describe some of the problems associated with providing them.

THE POTENTIAL AND PROBLEMS OF PUBLIC GOODS

Economists have strict definitions of public goods, but for our purposes it is probably sufficient to describe a public good as a product or service that is easily accessible to all people (it is difficult to exclude anyone from using it) and that can be used by many people at the same time (its use by one person does not preclude its use by any other person). Because the degree of accessibility and the degree of nonrivalry can vary, some public goods are more "pure" than others, but for simplicity we will ignore this distinction. In agriculture, examples of public goods include a high-yielding wheat variety, a labor-saving conservation tillage practice, a market information program broadcast over the radio, and public research in general (Winkelmann 1994), in fact any nonproprietary technology that is freely available to large numbers of people at little or no cost.

Although they may be highly desirable, public goods are not readily produced by profit-oriented private firms, because it is difficult for the producer of a public good to capture enough benefits to compensate the cost of production. To avert so-called "market failure," governments usually provide public goods because it is agreed to be in the interest of society. The government of India has invested heavily in agricultural research and extension, for example, to improve agricultural production and eliminate the famines that once ravaged the subcontinent. The government stepped in for a number of reasons, including the fact that private companies lack incentives to invest in a large R & D system to produce improved crop varieties that many farmers are too poor to buy. Even if most Indian farmers could afford to buy improved seed, many may choose not to, since they can easily acquire a small supply from a friend or neighbor and multiply it up on their own. Private firms are understandably reluctant to invest in the provision of products or services from which many individuals can be expected to benefit without helping to pay for the cost (a problem that economists term "free riding"). In summary, public goods are goods from which the supplier has difficulty in directly recovering investment costs and earning profits. Difficulty in recovering investment costs and earning profits does not mean that the benefits generated by investing in public goods are small, however. On the contrary, the benefits of public goods may be enormous, even though this may not be readily apparent when they are spread across a large number of beneficiaries. In India, for example, hundreds of millions of people now have access to more food at lower prices, and a major famine has not occurred in many years.

The issues of who pays the costs of public goods and what quantity of public goods is appropriate are contentious ones. In the case of agricultural research that is targeted at the poorest of the poor, those who pay the cost— typically taxpayers in the medium- and high-income brackets—often receive a relatively small portion of the benefits. At the same time, the main benefici-aries—peasant farmers and the urban poor who spend a large proportion of their incomes on food—may not have a say in deciding how government pri-orities are established. For these reasons, agricultural research is often funded at socially suboptimal levels.

The provision of public goods, including agricultural research, becomes more complex as the number of supplying organizations increases and as their constituencies become more diverse. In many countries, including the United States, every state or province has its own agricultural research organization, which is funded by local taxes and responsive to local needs—and therefore more likely to place local interests above national interests. Although certain types of research can be expected to produce large benefits at the national level, individual states lack incentives to carry out this research, because a large portion of the benefits can be expected to "spill over" to other states that are not sharing the costs. Recognizing that this is the case, national governments generally establish national research organiza-tions to provide goods and services that are deemed necessary for all but which will not necessarily be provided at the local level (for example, national standards for grading agricultural products).

Compared to the provision of national public goods, the provision of international (or global) public goods is an even more daunting prospect. As Kindleberger (1986) observes, "The tendency for public goods to be under produced is serious enough within a nation bounded by some sort of social contract, and directed in public matters by a government with the power to impose and collect taxes. It is . . . a more serious problem in international political and economic relations in the absence of international govern-ment." Peterson (2000) agrees that it is problematic for institutions to pro-vide global public goods in optimal amounts because no international authority exists to support agreements, but he points out that "international regimes" may compensate to some extent. These regimes can be thought of as "institutional structures designed to solve particular public goods problems in the absence of a world government" (Peterson 2000).

In this chapter, we propose to examine one particular global public goods problem: the provision of international research on improved maize and wheat technology to poor farmers in developing countries. In particular, we focus on how changes in the international environment are transforming the legal and social conventions (the "regime") governing international agricul-tural research and are affecting the flow of improved maize and wheat tech-nology to developing countries. We begin by describing the international research system that develops and delivers improved maize and wheat tech-nology for poor farmers. We then address several issues that will influence whether and how this system continues to operate. Aside from the public goods problems described earlier, these include such divisive issues as the rights to genetic resources and intellectual property protection. We describe strategies that the International Maize and Wheat Improvement Center (better known as "CIMMYT," its acronym in Spanish) has used to continue delivering global public goods to poor people, and conclude with a brief statement of why, despite the present uncertainty in the global research outlook, there will still be a place for international agricultural research and its products.

RESEARCH AND DEVELOPMENT FOR GLOBAL PUBLIC GOODS: ORIGINS AND ACHIEVEMENTS

In the two decades after World War II, changing perceptions about the potential socioeconomic effects of applied science led to a realization of the crucial role that research could play in hastening the economic development of poor countries. As noted by Press and Washburn (2000), the role of university research in developing technologies that altered the course of the war (penicillin and streptomycin as well as nuclear weapons) heightened government awareness that "academics were uniquely capable of undertaking crucial research initiatives," and public funding for research grew rapidly.

During the same period, European nations divested themselves of their colonies and protectorates, leaving a number of newly independent nations to make their way toward economic development with limited resources. By the mid-1960s, a number of these new nations seemed to be faltering on the path to prosperity. The prospect of famine and unrest in developing countries, especially in Asia, underscored the need for a new kind of development assistance. As it became increasingly clear that industrial development did not necessarily lead to sustained economic growth, "agriculture was gaining ascendancy in the economic strategies of developing countries" (Baum 1986). Confidence in the efficacy of publicly sponsored research began to merge with the conviction that agricultural research could stabilize and strengthen the economies of developing nations.

The case for agricultural research as a means of fostering economic growth in the developing world was compelling. The Asian subcontinent had been on the brink of famine in the mid-1960s, but in an exceptional international effort, researchers had developed new wheat and rice varieties that yielded much more than the wheat and rice varieties that Asian farmers already grew. When these new varieties were grown in India, Pakistan, and Bangladesh, they produced enough grain to make the difference between life and death for millions of people. The scale of this achievement, termed the "Green Revolution," was so widely recognized that Norman E. Borlaug, the plant breeder who had developed the wheat varieties, was awarded the Nobel Peace Prize in 1970. Governments, development banks, international organizations, and private foundations became convinced that funding international agricultural research—which efficiently produced technology that could be used in a wide number of developing countries—could have enormous international benefits, both economic and social. These organizations established a consortium, the Consultative Group on International Agricultural Research (CGIAR), to support agricultural research in developing countries. This consortium funded what have come to be called the Future Harvest Centers of the CGIAR.[1]

One of the first research centers in this consortium was the institute in which Norman Borlaug had conducted most of his research. Established in 1966, the International Maize and Wheat Improvement Center remains committed to improving the productivity, profitability, and sustainability of maize- and wheat-based cropping systems in developing countries. To a great extent, CIMMYT and the other Future Harvest Centers were modeled on the plant science departments of U.S. universities, especially the publicly funded land-grant universities, which had been highly successful in employing science to improve agriculture. The difference was that CIMMYT's mission was international, and to meet the needs of developing countries, the center emphasized applied rather than basic research. Like many land-grant universities, the center also had a role in educating researchers by providing highly specialized training to thousands of scientists from developing countries.

As it completes its thirty-fifth year, CIMMYT has many achievements to its credit. CIMMYT-related wheat varieties are planted on more than 64 million hectares in developing countries (more than three-fourths of the area planted to modern wheat varieties in those countries). CIMMYT-related maize varieties cover nearly 15 million hectares in nontemperate environments of developing countries (almost half of the area planted to modern maize varieties in those environments). Genetic diversity and the conservation of maize and wheat genetic resources have greatly improved. Innovative crop management practices designed to reduce environmental degradation and conserve resources have been developed, and farmers are using them.

All of these products and services originated in an extensive, international, collaborative research network that relied on the open exchange of knowledge and technology in the public domain. This network presently involves R & D organizations in more than one hundred developing nations, as well as similar institutions in numerous industrialized nations.[2] The benefits of collaboration between CIMMYT and national research organizations in developing countries have been impressive. Depending on several economic and technical assumptions, estimates of the value of the additional grain production attributable to international wheat breeding range from $2 billion to $4 billion per year (Cassaday et al., forthcoming). In the case of maize, the calculations are more complicated, mainly because of the large size of the private maize-breeding industry, but the economic value of the additional grain production attributable to public international maize-breeding efforts is likely to exceed $2 billion per year (Morris, personal communication).

The social benefits of this research, though difficult to measure, are also likely to be large. Additional maize and wheat production alone certainly cannot be credited with reducing malnutrition in the developing world, but the prevalence of malnutrition in developing countries declined from 46.5 percent in 1970 to 31 percent in 1995 (Smith and Haddad 2000), partly because more food was available to more people. Food has certainly become more affordable for consumers. Between 1982 and 1995, real world wheat prices fell by 28 percent, and world maize prices dropped by 43 percent (Pinstrup-Andersen, Pandya-Lorch, and Rosegrant 1999). Cheaper food is important for poor people in developing countries, who spend 50 to 80 percent of their disposable income on food, compared to the 10 to 15 percent spent in Europe and the United States (Pinstrup-Andersen and Cohen 2001).

International agricultural research has also benefited the environment. By breeding plant varieties with genetic resistance to pests and diseases, public research organizations made farmers' use of harmful, expensive agrochemicals unnecessary.[3] It has been argued that by increasing agricultural productivity per unit of land, research has prevented farmers from cropping more ecologically fragile land and from invading forested areas. Recent calculations (Grace et al. 2000) indicate that if the developing world had attempted to meet its food requirements in 1995 without the improved varieties of food crops[4] developed since the Green Revolution, an additional 426 million hectares of cropped area would have been needed (a fivefold increase over cropped area in 1965). An even more important finding is that this land savings helped to reduce greenhouse gas emissions by 35 percent. Grace and colleagues (2000) conclude that "without the Green Revolution . . . the atmospheric concentration of greenhouse gases would be significantly higher than they are at present and the actual onset of climate change may have hastened."

Many people would agree that those benefits are impressive and that it is a good thing that public-sector research delivers them. Even so, it is becoming increasingly challenging to provide international agricultural research to poor countries. The environment in which CIMMYT seeks to fulfill its mission is changing radically. The international community faces several choices that will greatly influence support for international agricultural research and the conditions under which that research will take place. In the following sections we explore those choices. Some of them are related to issues that typically surround the provision of public goods, whereas others reflect concerns over the mission and responsibilities of public institutions dedicated to development.

GLOBAL PUBLIC RESEARCH FOR POOR PEOPLE: THE CHANGING CONTEXT

Development strategies must adapt to a new constellation of circumstances that influence how international agricultural research will be conducted and how research products will be delivered in the future. These include declining investment in public research and international aid for agriculture; increasing complexity of the agricultural problems that science is being asked to solve; growing dissent over the conservation, exchange, and use of genetic resources; the proliferation of intellectual property rights and proprietary agricultural technologies; the rise of biotechnology and the associated controversy over genetically modified organisms; increasing economic and political power of the private sector; rising pressure for public-sector institutions to behave like private institutions; and growing concerns over scientific and social equity. Because these issues are so closely related, it is difficult to treat them separately, but we shall discuss each of them in turn, emphasizing the choices and dilemmas that they present to the international community.

DECLINING INVESTMENT IN INTERNATIONAL AGRICULTURAL RESEARCH

Between 1991 and 1996, development assistance fell by nearly 15 percent. From 1986 to 1996, development assistance directed specifically at agriculture fell almost 50 percent in real terms (Pinstrup-Andersen and Cohen 1998).[5] Much of the reduction occurred as the seven wealthiest countries that provided development assistance reduced their contributions. At the same time, many developing countries have reduced their own public spending on agriculture, partly under pressure from donor and lending institutions. The results of public underinvestment in agriculture may already be apparent. Pinstrup-Andersen and Cohen (1998) have found that, with the exception of China and India, in most low-income countries, agriculture grew by less than 3 percent per year from 1990 to 1996—not enough to keep pace with population growth.

INCREASING COMPLEXITY OF RESEARCH CHALLENGES

At the same time that the international community's disenchantment with agricultural research is affecting low-income countries, those same countries are facing agricultural challenges that most wealthy nations would find difficult to overcome.

Twenty years from now, the world's farmers will have to produce 40 percent more grain to meet demand for cereals, including wheat and maize (Pinstrup-Andersen, Pandya-Lorch, and Rosegrant 1999). In developing countries, the demand for wheat and maize will rise faster than demand for rice, the other major food staple.[6] In two decades, 67 percent of the world's wheat consumption and 57 percent of the world's maize consumption will occur in developing countries. Even with projected production increases, by 2020 wheat will constitute more than 50 percent of the developing world's net cereal imports. Maize will constitute 33 percent (Rosegrant et al. 1997).

Nearly everyone is aware that the world produces enough food to feed all of its people, but the challenge of supplying food to those who need it most is not as simple as it would appear. Simply increasing the "pile of food," observes Falcon (2000), "is by no means sufficient to assure food security among the poor. If developing countries with a large percentage of undernourished people are to solve employment, income, and food-access problems, most of the increased agricultural output must be grown within the borders of these nations."

That is where the challenge becomes acute, because the agricultural problems of developing nations are extreme. They range from numerous physical problems (such as diseases and pests, drought, floods, severe environmental degradation, and infertile soils) to institutional problems (such as weak extension programs, research organizations literally immobilized by lack of funding, limited access to agricultural inputs and credit, poor infrastructure, and poorly developed markets), to larger deficiencies in human and financial capital.

These challenges to agriculture are further complicated by the fact that agriculture itself is more complex. In the 1970s, agricultural research, whether it was international, national, or local, tended to be organized along commodity lines, focusing on specific, well-defined problems (e.g., breeding for higher yield, disease resistance, pest resistance; determining optimal fertilizer application levels). During the 1980s, the focus of research shifted gradually to cropping systems, which tend to be characterized by problems involving interactions between large numbers of on-farm and off-farm enterprises. To respond to these problems, research organizations shifted from monodisciplinary research to multidisciplinary research. By the 1990s, more researchers recognized that they needed to give greater attention to environmental and sustainability issues alongside the more traditional emphasis on productivity. At the beginning of the new century, research organizations are also being asked to demonstrate the linkages between technology development and poverty alleviation, which implies focusing more attention on the role of policies and institutions in fostering positive agricultural change.

Clearly, no single institution or technology will meet all of these research goals. Partnerships and consortia are essential for assembling the human expertise, technology, and often the financial capacity needed to make a difference for poor farmers. The complexity of the institutional arrangements supporting international research is growing. Organizations need time to explore, access, and assemble promising research tools and technology. They need time to assess which organizations would be effective research partners. The large number of partners in international research efforts—including funding agencies, nongovernmental organizations (NGOs), private companies, and farmers (through participatory research)—makes it harder to reach agreement on how best to operate. Transaction costs increase, and partners have to establish clear guidelines and decision rules to govern their collaboration. The large effort to secure international collaboration may certainly be worthwhile, however (witness the results of CIMMYT's maize and wheat research). As Peterson (2000) points out,

> In addition to the benefits of international public goods that would not be supplied in the absence of international cooperation, international organizations may generate other benefits for participants. Efficiency gains due to scale economies in the provision of the public good, the greater amount of information made available through the supranational structures, and increased political prestige for those who participate in the agreement are examples. As with costs, these benefits increase with the number of participants and the degree of integration.

DISSENT OVER GENETIC RESOURCES

Herdt (1999) describes the controversy over the exchange, use, and control of plant genetic resources as the enclosing of the "global genetic commons." He notes that "changing technology and institutions have interacted throughout history to create property rights from what had previously been public goods," and the ability to manipulate DNA has "generated a new class of asset whose ownership is now being contested by multi-billion-dollar companies." As private companies have increasingly obtained intellectual property rights to plant traits, genes, and very small genetic components, many agencies in developing countries have come to believe that their genetic resources may prove potentially valuable to the emerging biotechnology industry. Angered by what they consider "biopiracy," and concerned that they may one day be denied access to what they consider their own resources, many countries in which genetic resources have been collected are demonstrating less willingness to make genetic resources freely available for use by others.

Many issues related to the conservation, ownership, and exchange of genetic resources remain to be resolved at the national and international levels.[7] The net effect of the trends we have just described, however, has been to reduce the flow of genetic resources for research and create a great deal of uncertainty in the public sector about how to work in an environment where the rules of the game are changing, perhaps beyond recognition. Cassaday et al. (forthcoming) observe that the rules established through the international negotiations on plant genetic resources for food and agriculture, which are under way at the Food and Agriculture Organization (FAO), "could dramatically affect the origins, i.e., the genetic content" of all crops "vital to food security and economic development." They conclude that if the negotiations fail to develop a system that is conducive to international public plant breeding, "governments will have to be prepared to devote considerably greater financial and human resources to plant breeding and acquisition of materials than they seem prepared to provide today." In other words, the commitment to provide a particular set of public goods previously provided through international channels will shift to national governments working individually, which is likely to be a less efficient alternative.

THE RISE OF INTELLECTUAL PROPERTY RIGHTS AND PROPRIETARY TECHNOLOGY

Preston (2000), reviewing trends and achievements at the U.S. Patent and Trademark Office during the Clinton administration, reported that since 1993,

> patent and trademark filings have increased more than 70%. Patent filings have gone from 174,000 in 1993 to just under 300,000 this year, and trademark applications have increased from 140,000 to over 370,000. The sheer volume of all of this data has won the USPTO the distinction of having more data storage than the combined contents of every book in the Library of Congress.

Obviously not all of these patent filings were related to agricultural research, but it is certain that applied agricultural research in biotechnology generated a good number of them. The proliferation of patents and other forms of intellectual property protection could possibly spread beyond the United States, because many countries are required to adopt intellectual property regimes as part of world trade agreements (Morris and Ekasingh, forthcoming).

As noted, the private sector traditionally did not invest in developing new plant varieties, mainly owing to the nonappropriability of benefits. Private investment in agricultural R & D increased in the 1930s and 1940s with the advent of hybrid maize seed companies (since the nature of hybrids is that the benefits become appropriable). In the 1990s, as the potential of biotechnology became clear, the "business of breeding" really began. Private companies poured money into R & D. Some observers believe that in the United States, the private sector's expenditure on research now exceeds expenditures by the public sector.

What motivated the rise in private-sector investment in plant breeding? Without a doubt, the potential for claiming intellectual property rights to plant varieties, genes, alleles, and other genetic components has driven this investment. Because the techniques of molecular genetics made it possible to identify the developer of a plant variety without question, it became far easier to claim intellectual property over plant varieties (Herdt 1999). In other words, plant variety protection (PVP), patents, and other types of intellectual property rights have made it possible for companies to appropriate benefits from investments in plant breeding, thereby converting what was once a public good into a private good.

Not everybody is comfortable with this development. As genetic resources in their many forms—plant varieties, the genetic components of plants, and the information associated with them—have gained in value, they are increasingly perceived to be strategic assets, and many observers are disturbed to see that the private sector is appropriating the rights to these assets. Critics are especially concerned by what they see as a fairness issue—many of the genetic resources used as inputs into modern breeding programs were improved by farmers over thousands of years of on-farm selection, and it is not obvious that these farmers (or their descendents) are being compensated.

Even though the new appropriability of genetic resources provides an incentive for private companies to invest in research, another fairness issue emerges when one considers that private companies develop products only for commercial markets. They rarely develop products for the many poor farmers, especially subsistence farmers, who cannot afford to pay for them.

The drive toward intellectual property rights has obviously changed the ethos of plant breeding in the public sector, which relied on "a willingness to share discoveries and materials for the common good" (Herdt 1999). Presently, public-sector scientists developing research products for poor people find it increasingly difficult and costly to access the products and processes required for their research. Because researchers may have to work with a number of patented "enabling technologies" to achieve their goals (these technologies can include molecular constructs, transformation processes, genes, and traits), the amount of time spent negotiating access to technology is likely to erode the time and money spent developing and delivering it.[8] Furthermore, one of the most difficult choices facing public organizations is whether to seek intellectual property protection over their own products. Their motivation is not so much to profit from this action as to prevent other agencies from appropriating rights to their research materials and making them difficult and/or expensive for others to access. Falcon (2000) concludes that "the fear that 'outsiders' will patent existing products . . . has left national agricultural research systems and the international agricultural research centers in a quandary as to whether or not to employ patenting as a defensive strategy against bio-piracy."

Finally, another implication of the rise of proprietary agricultural technology is that researchers in the public arena no longer face a simple decision about *which technology* to use in their research. Because property rights link a technology with its owner, researchers more often than not are also deciding which corporate entity they must partner with—or pay—to achieve their goals. Although many observers worry that alliances with private organizations are nothing less than exploitative, others believe that the only realistic strategy is to build alliances that achieve public goals, even if they also benefit private bank accounts.

DISSENT OVER BIOTECHNOLOGY

Widely differing perceptions about the potential benefits and drawbacks of biotechnology have colored an active and very public debate that covers a range of scientific, political, and ethical issues. Tripp (2000) observes that proponents of biotechnology "argue for the need to increase food production and point to the possibility of addressing the problems of marginalized farmers," whereas opponents "question the safety, relevance, and equity of the new technology." Some have gone so far as to mandate a moratorium on the release of genetically modified organisms (GMOs) and a complete cessation of research. Others have sought to ensure that the views of developing countries are represented in this debate, fearing that the most food-secure nations will make decisions with repercussions for the least food-secure nations. Organizations concerned with agricultural research are now deciding where they stand on these issues.

For international research organizations such as CIMMYT, one problem resulting from this dissent—which is fuelled in part by incomplete knowledge and false information—is that it threatens to close off many avenues of potentially productive research for developing countries. Pinstrup-Andersen and Cohen (2001), in an extensive review of this problem, observe that "positions for or against the use of genetic engineering in food and agriculture in industrialized countries are frequently extrapolated directly to developing countries," even though the perspectives and interests of groups in industrialized and developing nations differ greatly with respect to the technology. For example, rich nations can afford to worry about the health consequences of genetically modified (GM) food and determine that it is better to abandon research and commercialization of GM food crops, whereas many poor nations may find it in their interest to explore GMOs' potential for increasing food production and agricultural export earnings. Research on GM food crops may diminish if industrialized nations decide that it is safer to use GM technology to develop pharmaceuticals, with the result that many technologies of potential use for agriculture in developing countries will not be developed. Falcon (2000) comments that developing countries "express concern that key research initiatives with biotechnology will not be pursued because of what they perceive to be the private sector's focus on the wrong products, for the wrong reasons, at the wrong time." Pinstrup-Andersen and Cohen (2001) note that these divergent views are likely to lead rich and poor nations to adopt different policies and standards that "may conflict with the current globalization trends," and that "for globalization to continue in the area of food and agriculture, certain policies and standards need to be synchronized, and the biggest threat is that low-income countries will have to adopt policies and standards that are appropriate only for high-income situations."

THE PREDOMINANCE OF THE PRIVATE SECTOR

As noted, private organizations have marshaled an impressive array of financial and human resources to support their agricultural R & D goals. Heisey, Srinivasan, and Thirtle (2000) have assessed investments by the public and private sector in plant-breeding research in several settings (Australia, Canada, the UK, and the United States). They found that "across industrialized countries and across crops, the general trend has been towards relatively greater private sector investment in plant breeding, and greater use of private sector varieties in farmers' fields." In the United States, "it is likely that for field crops alone private plant breeding expenditures now surpass public expenditures by a considerable margin."[9] Morris (forthcoming) documents that in developing countries, the private sector now invests more in maize breeding than the public sector.

Biotechnology research especially highlights the contrast between public and private investments in agricultural R & D. According to Sandburg (1999), the National Science Foundation provided $30 million for plant genomics research in 1998 and $50 million in 1999, whereas private companies spent $1.5 to 2 billion. When the Novartis Agricultural Discovery Institute Inc. (NADII, now known as the Torrey Mesa Research Institute) was established in California in 1998, funding for the first ten years was anticipated to be $600 million. (Funding for all seventeen Future Harvest Centers of the CGIAR, by comparison, was $340 million in 1998.) The financial clout of the multibillion-dollar biotechnology industry, which had its origins in technology developed in the public sector, has come to have implications for how the public sector—including organizations such as CIMMYT—chooses to do its work. Some of these implications are discussed in the next section.

PRESSURES FOR THE PUBLIC SECTOR TO ACT LIKE THE PRIVATE SECTOR

The private sector traditionally has supported public-sector research in many ways, including direct research grants, donations, endowed chairs, and scholarships. More recently, however, private organizations seem to be financing a greater share of the research in universities[10] and public organizations under arrangements that have called the independence of the public sector into question.

In 1998, NADII reached a still-controversial agreement with the University of California at Berkeley in which NADII agreed to provide $25 million over five years to the university's Department of Plant and Microbial Biology to conduct basic research on plant genomics. Under the conditions of the grant, department researchers do not work on specific products for Novartis (now renamed Syngenta), but Syngenta receives the first right to negotiate licenses on about one-third of the department's discoveries—discoveries from research funded by NADII/Torrey Mesa as well as by state and federal organizations. Syngenta benefits from gaining access to research that could yield commercial products, and the university gains access to Syngenta's proprietary gene-sequencing database, an immensely valuable resource otherwise unavailable to the university. A committee formed by three professors from the department and two Syngenta representatives decides which research projects to fund through the Torrey Mesa grant. The decision to award Syngenta the right to negotiate for 30 to 40 percent of the department's inventions was based on the fact that the Torrey Mesa grant would fund 30 to 40 percent of the department's annual research budget (Sandburg 1999).[11]

Many observers of events at UC Berkeley felt that this arrangement blatantly challenged the university's mission as a public institution committed to preserving academic freedom—particularly the freedom to ensure that its research agenda remained independent of commercial interests. Others wondered how much longer the university could claim to serve the public good in any case, given that state funding for UC Berkeley had fallen to 34 percent in 2000, compared to 50 percent twelve years previously (Press and Washburn 2000).

Similar doubts have been expressed in response to the trend among universities and other public organizations to patent their inventions. Universities that once regarded patents as "fundamentally at odds with their obligation to disseminate knowledge as widely as possible" have altered their way of doing business so that "nearly every research university in the [United States] has a technology-licensing office" (Press and Washburn 2000). Many university campuses are now surrounded by clutches of start-up companies formed on the basis of university discoveries. Despite a handful of lucrative successes,[12] however, most licensing offices have yet to become a major source of income for the universities they serve (Press and Washburn 2000).

Strategies such as these have raised questions that have echoed throughout the public sector. Some of these questions are related to the future of public-sector research itself, whereas others are related to the increasingly blurry distinction between public and private research. If the public sector is unwilling to increase funding for research, will public research organizations continue to achieve their goals? Many fear that if public institutions cannot compete with the resources offered by the private sector, they will no longer attract and retain the best researchers. Others worry that public organizations will not conduct the kind of basic research that truly inspires innovation. Still others have expressed great concern that the public research agenda, which often addresses issues and meets needs of little importance to the private sector, will become distorted by the private sector's goals. Can intellectual freedom be protected if private rather than public funding increasingly supports public institutions? Finally, what really defines a "public" institution? Public institutions may profess to serve the public good, but will their financial statements and research portfolios give the lie to that assertion? We will return to some of these questions, and to how CIMMYT and other international research organizations are attempting to deal with them, later in this chapter.[13]

THE STRUGGLE FOR EQUITY IN SCIENCE

Diverging investment in agricultural R & D by the public and private sectors, diverging perspectives in wealthy and poor nations about potential applications of biotechnology, concerns over the ability of public organizations to access new technologies and processes for research, and the complexity of the new technology itself have raised the twin specters of "scientific imperialism" and "scientific apartheid." We have already discussed fears that wealthy nations will dictate the scientific limits of poor nations, not only through the kinds of research they choose to undertake but through the positions advocated by civil society organizations and development assistance agencies. Pinstrup-Anderson and Cohen (2001), for example, have shown how opposition to biotechnology in developing countries by civil society organizations in industrialized nations has elicited the response from some developing nations that they would prefer to determine for themselves, on the basis of their needs and values, whether and how they will use GMOs and other products of biotechnology. Herdt (1999) cautions that the increasing use of intellectual property rights could "raise costs or discourage innovations in the developing world, or shift power unfairly to industrialized country firms away from developing country organizations."

A parallel concern is that results of research undertaken in industrialized nations will be increasingly beyond the reach of most developing nations, and that the technology gap will only become wider over time. With some notable exceptions,[14] most developing nations lack the financial and human resources to mount ambitious biotechnology research programs, either publicly or privately funded. Nor do many nations have the resources to access technology that has already been developed (including GMOs) and monitor its use. Serageldin and Persley (2000) state the problem simply: "The economic concentration of investment, science, and infrastructure in industrial countries and the lack of access to the resulting technologies are major impediments to the successful applications of modern biotechnology to the needs of global food security and to create wealth for the presently poor people and countries."

THE STRUGGLE FOR SOCIAL EQUITY

If international research organizations such as CIMMYT are preoccupied by the prospect of growing inequity in science, it is because they are even more preoccupied by the prospect of growing inequity in society. One of the arguments in favor of international agricultural research is that its benefits are felt by the poorest members of society. Who are these people whose lives will be affected if agricultural research fails to help them? They include the nearly three billion individuals who survive on less than two dollars per day—two dollars for food, clothing, shelter, education, medical treatment, and other needs. They include the world's 160 million malnourished children. They include the people who live in rural areas in developing countries and depend on agriculture to provide income and food security—more than 70 percent of the population.

Should the interrelated trends discussed previously combine to inhibit international public agricultural research, fewer research organizations may survive to act as "agencies for equity." Although international agricultural research seems to be hemmed about with a growing number of constraints, at CIMMYT we believe that these constraints are not insurmountable. Our strategies for ensuring that international research empowers poor people and eradicates scientific apartheid—in short, our strategies for keeping public goods *public*—are discussed next.

CIMMYT: FREEDOM TO ACHIEVE A GLOBAL HUMANITARIAN MISSION

In adapting its research strategies to a volatile new research environment, CIMMYT is fully aware that "society benefits when the public sector has 'freedom to operate,' when it maintains public access to research tools subject to intellectual property protection by the private sector, and when it engages in fruitful collaborative research" (Heisey, Srinivasan, and Thirtle 2000). Here we outline some of the alternatives that will ensure that CIMMYT remains effective and true to its mission in the midst of great change in the way research is conducted.

PARTNERSHIPS FOR A NEW RESEARCH ENVIRONMENT

Especially as a result of new intellectual property arrangements, public research organizations are entering into a larger variety of research partnerships than in the past. These partnerships range from traditional philanthropic arrangements to purely commercial alliances and include direct support for research, collaborative public-sector research, licensing (different agreements for sharing costs and technology), market segmentation, technology grants for research in developing nations, and joint ventures (Falcon 2000). Here we will not discuss CIMMYT's more "traditional" research partnerships (although these, too, are changing as they come to involve a wider range of participants).[15] Instead we will describe (1) alliances between CIMMYT and the private sector and (2) partnerships between CIMMYT and other public research organizations in which processes or products of biotechnology are used.

Partnerships with Private Research Organizations

Private and public research organizations increasingly agree that it is urgent to explore the ways that their interests may intersect for the benefit of society. For example, the development of drugs to combat AIDS and the breakthrough with golden rice have raised awareness that private corporations may have a moral responsibility to make their products available to poor nations (in other words, under certain conditions, the private research sector should further the objectives of the public sector).[16]

   With regard to agreements with the private sector, CIMMYT's policy is to enter into such agreements only if they enhance the center's ability to achieve its mandate of service to the resource poor and the environment. In simple terms, will an agreement help CIMMYT to more quickly develop new, appropriate technologies and deliver them to farmers' fields in developing countries? If so, the agreement is what we call a "win-win" alliance, and the center can participate. Within this framework, CIMMYT has established four agreements with private research organizations that provide access to expertise and technologies that otherwise would not be available.[17] Three of the agreements involve research on wheat: a project to evaluate the potential of hybrid wheat; a project with a private company in

Spain to improve disease resistance, yield, and quality in durum and bread wheat; and a project with a private company in Mexico to improve the industrial quality of bread wheat.[18]

A fourth project aims to develop apomictic maize plants. Since 1990, a joint project between CIMMYT and France's Institut de Recherche pour le Développement (IRD), a public research organization, has focused on understanding apomixis (the asexual reproduction of plants through seed) and how the trait might be transferred to maize. To accelerate progress in this potentially revolutionary area, in 1999 CIMMYT and IRD formally entered into an important research collaboration with three private seed companies (Pioneer Hi-Bred International, Groupe Limagrain, and Syngenta). The five-year agreement is aimed at further understanding apomixis, which is the natural ability of some plants to reproduce offspring identical to the mother plant through asexual reproduction.[19] In the plant kingdom, more than four hundred species, most with little or no agronomic potential, possess this apomictic characteristic. Greater knowledge about this natural plant mechanism could provide the basis for its transfer to some of the most commonly grown agricultural crops, for instance, hybrid maize. For the agreement's seed-producing partners, enhanced knowledge of apomixis might create new options for improved multiplication and quality of seeds. For CIMMYT and IRD, the transfer of apomixis to maize offers the long-term possibility of delivering superior hybrid crop traits such as disease resistance and higher yields to the resource-poor farmers of the world through the inherent reproductive characteristics of apomictic plants.

Partnerships between Public Research Organizations to Explore the Potential of Biotechnology

CIMMYT conducts a wide range of biotechnology research and collaborates with a number of public-sector institutions in developing countries. Here we will describe a project recently initiated in Kenya. Scientists from the Kenya Agricultural Research Institute (KARI) and CIMMYT are using conventional as well as biotechnological breeding strategies to develop maize resistant to stem borers, which are estimated to destroy 15 to 40 percent of Kenya's maize crop each year. The Insect Resistant Maize for Africa (IRMA) Project is funded by the Novartis Foundation for Sustainable Development. The project was launched through a consultative meeting in which all groups concerned with the outcome met to discuss their views of the project, including representatives from KARI and CIMMYT as well as from farmers', women's, and church associations; extension services; various government ministries; the private sector; and a contingent of Kenyan print and broadcast media.

Over five years, researchers participating in the IRMA Project will develop integrated pest management strategies and use conventional and biotechnological means (including resistance based on Bt genes) to breed insect-resistant maize for major Kenyan production systems and insect pests. The project will also establish procedures to provide insect-resistant maize to resource-poor farmers; assess the impact of insect-resistant maize in Kenyan agricultural systems; transfer skills and technologies to Kenya to develop, evaluate, disseminate, and monitor insect-resistant maize; and plan, monitor, and document the project's processes and achievements for dissemination to other developing countries, particularly in East Africa.

The project calls on CIMMYT and KARI expertise in maize breeding, agricultural economics, biotechnology, entomology, and communications. It is important to emphasize that project researchers have agreed to identify and develop gene constructs that contain no herbicide or antibiotic markers. Maize varieties produced by the IRMA Project will carry only "clean" or "purified" Bt genes, circumventing concerns about unforeseen impacts on the environment or human health. While this approach costs more and takes longer, IRMA researchers are committed to addressing all reasonable issues that emerge regarding the technology.

POLICIES FOR A NEW RESEARCH ENVIRONMENT

As the research environment becomes more complex and public research organizations enter into more alliances with the private sector, CIMMYT has sought to develop clear, open policies with respect to intellectual property and new technology. These policies provide a public account of CIMMYT's strategies for making its research products available to the international community and for working with private organizations in ways that are absolutely consistent with its mission to help the poor.

CIMMYT'S INTELLECTUAL PROPERTY POLICY AND INTELLECTUAL PROPERTY MANAGEMENT

In 2000, CIMMYT released its policy on intellectual property (CIMMYT 2000a).[20] The preamble emphasizes the center's concern over preserving public access to its research products:

> As a publicly funded international research institute, CIMMYT regards its research products as international public goods. Yet, in the current political and legal environment, producing and keeping the products of its research in the public domain, free for use and development both by scientists and farmers, have become increasingly problematic. It is in this context that CIMMYT has examined, and will continue to examine, its policies and practices in regards to intellectual property rights.

For the most part, the policy spells out procedures for managing intellectual property that were already in place within CIMMYT. One example is the procedure to hold the genetic resources designated under a 1994 FAO/CGIAR agreement in trust for the benefit of the international community, especially developing countries. The new intellectual property policy represents a departure from previous modes of operation, however, by establishing that CIMMYT will take steps to protect its inventions through patents, plant breeders' rights, copyrights, trademarks, statutory invention registrations or their equivalent, and/or trade secrets under the following conditions:

> To support public and private partnerships which pursue mission-based research or which develop and apply research results; to assure ready access by others to research products developed or funded by CIMMYT; to avoid possible restrictions arising from "blocking" patents and to ensure CIMMYT's ability to pursue its research without undue hindrance; to facilitate the transfer of technology, research products, and other benefits to the resource poor including, where appropriate, through commercialization or utilization of research products; and/or to facilitate the negotiation and conclusion of agreements for access to proprietary technologies of use to CIMMYT's research and in furtherance of its mission.

The policy further specifies that, in light of the "evolving legal and political environment," CIMMYT's board of trustees will "regularly review this Policy and its implementation in order to ensure that CIMMYT is well positioned to carry out its mission."

CIMMYT was one of the first Future Harvest Centers to release an intellectual property policy, and the press quickly noted the decision. The policy was described in *Nature,* which quoted a CIMMYT board member as saying that "this is not an effort by the organization to 'get rich' by patenting discoveries, but to ensure broad distribution of plant materials through a flexible policy" (Dalton 2000).

Aside from its new policy, CIMMYT has several organizational avenues for managing intellectual property and related issues. For a number of years, CIMMYT has maintained standing committees on intellectual property, biosafety, and bioethics, and it has conducted a full intellectual property audit. In 2001, an Intellectual Property Management Unit was established to provide further guidance and leadership on intellectual property issues.

CIMMYT's Genetic Engineering Strategy

In developing the tenets of its genetic engineering strategy for wheat and maize, CIMMYT has emphasized the needs of its partners in national research organizations and the usefulness and safety of its products for farmers. Five points guide the center's genetic engineering program:

- Plant varieties that are genetically engineered by CIMMYT are developed in concert with a national program partner to meet a delineated need.
- CIMMYT provides only transformed plants that carry "clean" events, meaning that only the gene of interest is inserted into the final product.
- No transformed plants that carry selectable markers, such as herbicide or antibiotic resistance, are provided to national programs for release.
- CIMMYT's focus on possible genes for transfer is only on plant, bacterial, fungal, and viral genes (i.e., not on animal genes, especially human genes).
- CIMMYT works only in countries that have biosafety legislation and regulations.

NEW MECHANISMS FOR ACCESSING AND PROVIDING RESEARCH PRODUCTS

CIMMYT actively seeks new ways of accessing research tools and providing research products. It has been suggested that one potential means of reducing some of the complicated legal arrangements involved in accessing and disseminating new technology is a mechanism known as a "patent pool," which has been used in the United States for 150 years, mostly in the manufacturing industry and more recently in the electronics industry (Clark et al. 2000).[21] This mechanism, which is regulated by the Antitrust Division of the U.S. Department of Justice and by the Federal Trade Commission, is thought to offer "one way to address the issue of access to vital patented biotechnology products and processes" (U.S. Patent and Trademark Office 2001).

A patent pool consists of two or more patent owners who agree to license one or more of their patents to one another or third parties. The pool has the advantage of allowing organizations to make all of the components needed to conduct a process or produce a technology available from one source. Ideally, according to Clark et al. (2000, 8–9), such arrangements would make it possible to integrate complementary technologies, reduce transactions costs, clear blocking positions, avoid costly infringement legislation, and promote the dissemination of technology. These authors note that "the re-emergence of . . . patent pools suggests that the social and economic benefits of such arrangements outweigh the costs." They contend that patent pools can encourage "the cooperative efforts needed to realize the true economic and social benefits of genomic inventions. In addition, since each party in a patent pool would benefit from the work of others, the members may focus on their core competencies, thus spurring innovation at a faster rate." Should patent pools become more common in the biotechnology industry, public research institutes may have greater access to technology for their research—if the conditions are acceptable to their goals as public institutions.

Increasingly, the international community appears to be seeking a forum for reconciling the objectives of private and public research institutions with respect to developing countries. A white paper issued by seven academies of science in 2000 advocates the establishment of an international advisory committee to "assess the interests of private companies and developing countries in the generation and use of transgenic plants to benefit the poor—not only to help resolve the intellectual property issues involved, but also to identify areas of common interest and opportunity between private sector and public sector institutions" (Anonymous 2000). Such an advisory group could become a valuable resource for international research organizations working in developing nations. The current Central Advisory Service on Intellectual Property of the Future Harvest Centers—an international forum that has already been established—could possibly play this role, at least within the CGIAR.

Meanwhile, private research organizations have become increasingly aware of the importance of collaborating with public research initiatives, especially in developing countries, and this trend could be beneficial for international agricultural research.[22] Nash (2000) reports that the developers of golden rice, Ingo Potrykus and Peter Beyer, "struck a deal with AstraZeneca, which . . . holds exclusive rights to one of the genes Potrykus and Beyer used to create golden rice. In exchange for commercial marketing rights in the US and other affluent markets, AstraZeneca agreed to lend its financial muscle and legal expertise to the cause of putting the seeds into the hands of poor farmers at no charge." In January 2001, Syngenta, one of the world's largest agrochemical companies, published the first complete genome of a food crop (rice). With this information, Syngenta researchers can achieve highly specific breeding objectives very rapidly, because it enables them to identify particular genes (such as a gene conferring resistance to an important disease) that would be useful to transfer from one variety to others. The value of this information for public breeding research in developing countries is obvious. Syngenta, in a statement issued on January 26, 2001, said that in developing countries, "where rice is a vital crop, Syngenta will work with local research institutes to explore how this information can best be used to find crop improvements to benefit subsistence farmers. It is our policy to provide such information and technology for use in products for subsistence farmers, without royalties or technology fees." Many members of the development community welcomed this effort to channel important technology to the people who arguably have most to benefit from it.

DIALOGUE, ADVOCACY, AND INFORMATION
FOR RESEARCH PLANNING

CIMMYT researchers are committed to conducting an open dialogue on many of the scientific, legal, and institutional changes that are transforming the environment in which it seeks to fulfill its mission. For example, CIMMYT convened an international forum in Tlaxcala, Mexico, in late 1999 to initiate a dialogue on key issues related to public-private alliances in agricultural research. The participants were all highly respected, experienced individuals active with the private sector, major public research institutes in the developing world, multilateral donor agencies, academia, and the CGIAR. Participants produced a statement that reflected their consensus on how public and private research organizations could adopt complementary and mutually reinforcing forms of working together (CIMMYT 2000b).[23] CIMMYT also participates in numerous international forums on biotechnology, including the Africa Biotechnology Stakeholders Forum, which addresses the special issues surrounding biotechnology in the African context.

Because it is a well-known research institution with decades of experience in developing countries, CIMMYT can also help educate a wider audience about issues that are important to agricultural development. The IRMA Project has a strong public education and awareness program related to the technology it develops, for example. This program is directed at the general public as well as farmers and other important groups who have an interest in the project's progress and outcomes. In addition, CIMMYT has publicly advocated that national governments—not the governments of industrialized nations—must make their own decisions with respect to GMOs and other products of biotechnology, and that those decisions should emerge from "serious discussion based on credible, science-based information" (Feldmann, Morris, and Hoisington 2000).

Through its own research, CIMMYT is careful to provide sound information for research management decisions related to new technology and the research environment. For example, a CIMMYT researcher and a colleague from a national public research program recently reviewed the ways that public and private plant-breeding organizations could reorient their efforts (Morris and Ekasingh, forthcoming). Concluding that "if public breeding programs do not change the way they operate, they will become marginalized," Morris and Ekasingh also caution that public breeding programs should not simply "withdraw completely from areas claimed by the private sector, because a continuing public-sector presence may be desirable for efficiency as well as equity reasons." They identify five "essential functions" for public breeding programs in the future: the conservation of genetic resources; training of plant breeders; varietal testing and evaluation;

biosafety regulation; and crop genetic improvement for carefully selected products, traits, and crops (e.g., those that are important for subsistence farmers).

CIMMYT researchers have also published research and review articles on gene flow between GMOs and other crops, the efficiency of conventional breeding methods compared to molecular breeding, economic and social incentives for farmers to preserve genetic resources in situ, economic returns on conserving and using genetic resources, flows of wheat and maize genetic resources between developing and developed countries, the implications of providing GM seed to farmers in developing countries, and many other topics.[24]

GLOBAL PUBLIC GOODS FOR POOR FARMERS: STILL A REALITY

Much of the recent investment in agricultural research has been made by the private sector to meet commercial needs and satisfy stakeholders, but it has not been used directly to generate global public goods. As private interests in agricultural research grow, what will be the fate of international agricultural research and the goods and services it provides? Will centers such as CIMMYT remain active forces for agricultural development, or will they merely represent an outmoded way of doing business?

We do not underestimate the challenges, but we strongly believe that international agricultural research will continue to be an effective force for change in the developing world. To some extent, the current fierce debate over intellectual property and other issues has obscured the fact that much of the history of agriculture, including agricultural research, consists of a series of upheavals and accommodations that occurred as public and private organizations sought to adapt to economic and institutional change (Heisey, Srinivasan, and Thirtle 2000). Although the research environment has become extremely volatile, it is to be hoped that the international community will mobilize its considerable authority and resources to support international research and ensure that poor people are not excluded from development opportunities through shortsighted policies, agreements, and purely commercial interests.

We know that if less technology is generated to help the poor, and if fewer poor people can access that technology, then the cost of social equity will rise for every individual on earth. The consequences for the rural poor will not be small. Some people will pay with hunger and poor health; others will take their suffering to overcrowded cities or across borders to wealthy nations. At CIMMYT, we believe it is urgent to join actively in the debates surrounding international agricultural research and discover more efficient ways of fulfilling our mission. The penalty for not acting, and for being excluded from new research opportunities, is going up exponentially—each year, and for each person.

REFERENCES

Anonymous. 2000. *Transgenic plants and world agriculture.* Washington, DC: National Academy Press.

Baum, W. C. 1986. Partners against hunger: Consultative Group on International Agricultural Research. Washington, DC: World Bank for the Consultative Group on International Agricultural Research.

Cassaday, K., M. Smale, C. Fowler, and P. W. Heisey. Forthcoming. Benefits from giving and receiving genetic resources: The case of wheat. *Plant Genetic Resources Newsletter.*

CIMMYT (International Maize and Wheat Improvement Center). 2000a. *International Maize and Wheat Improvement Center policy on intellectual property.* Mexico City: CIMMYT.

CIMMYT (International Maize and Wheat Improvement Center). 2000b. *Tlaxcala statement on public/private sector alliances in agricultural research: opportunities, mechanisms, and limits.* Mexico City: CIMMYT.

CIMMYT (International Maize and Wheat Improvement Center). 2000c. Transparency is important. In *Science and sustenance: Annual report, 1999–2000.* Mexico City: CIMMYT, 70.

Clark, J., J. Piccolo, B. Stanton, and K. Tyson, with M. Critharis and S. Kunin. 2000. *Patent pools: A solution to the problem of access in biotechnology patents?* Washington, DC: U.S. Patent and Trademark Office.

Dalton, R. 2000. Cereal gene bank accepts need for patents. *Nature* 404 (April 6): 534.

Dreher, K., M. L. Morris, M. Khairallah, J.-M. Ribaut, S. Pandey, and G. Srinivasan. 2000. Is marker-assisted selection cost-effective compared to conventional plant breeding methods? The case of quality protein maize. Paper presented at the 3rd Annual Conference of the International Consortium for Agricultural Biotechnology Research (ICABR), August 15–28, Ravello, Italy.

Dreher, K., M. L. Morris, J.-M. Ribaut, and M. Khairallah. 2001. Potential impacts of biotechnology-assisted selection methods on plant breeding programs in developing countries. Paper presented at the preconference workshop, "Agricultural Biotechnology: Markets and Policies in an International Setting," at the Annual Conference of the Australian Agricultural and Resource Economics Society, January 22, Adelaide, Australia.

Falcon, W. P. 2000. Globalizing germplasm: Barriers, benefits, and boundaries. Paper presented at the 24th International Conference of Agricultural Economists, August 13–18, Berlin, Germany.

Feldmann, M. P., M. L. Morris, and D. Hoisington. 2000. Genetically modified organisms: Why all the controversy? *Choices: The Magazine of Food, Farm, and Resource Issues,* first quarter, 8–12.

Grace, P., P. Sanchez, J. Ingram, C. Palm, Reiner Wassman, M. Fisher, R. Thomas, F. Chandler, W. Bowen, R. Reid, M. Wopereis, and S. Waddington. 2000. The consequences of international agricultural research on greenhouse gas emissions and global climate change. Unpublished report prepared for the Inter-Center Working Group on Climate Change, Consultative Group on International Agricultural Research (CGIAR), Washington, DC.

Heisey, P. W., C. S. Srinivasan, and C. Thirtle. 2000. Privatization of plant breeding in industrialized countries: Causes, consequences, and the public sector response. Paper presented at the mini-symposium, "Emerging Issues in Agricultural Research Policy and Funding: Lessons from Case Studies," at the 24th International Conference of Agricultural Economists, August 13–18, Berlin, Germany.

Herdt, R. W. 1999. Enclosing the global plant genetic commons. Paper presented at the China Center for Economic Research, May 24, Peking University, Beijing, China.

Kindleberger, C. P. 1986. International public goods without international government. *The American Economic Review* 76 (1): 1–13.

Morris, M. L. Forthcoming. *Impacts of international maize breeding research, 1966–97.* Mexico City: International Maize and Wheat Improvement Center (CIMMYT).

Morris, M. L., and B. Ekasingh. Forthcoming. Plant breeding research in developing countries: What roles for the public and private sectors? In *Agricultural Research Policy in an Era of Privatization: Experiences from the Developing World,* ed. D. Byerlee and R. Echeverría. Wallingford, UK: CAB International.

Nash, J. M. 2000. Grains of hope. *Time Magazine* 156 (5).

Paarlberg, R. 2000. The global food fight. *Foreign Affairs* 79 (3): 35–36. Cited in RAFI (2000).

Peterson, E. W. F. 2000. The design of supranational organizations for the provision of international public goods: Global environmental protection. *Review of Agricultural Economics* 22 (2): 355–69.

Pinstrup-Anderson, P., and M. J. Cohen. 1998. *Aid to developing country agriculture: Investing in poverty reduction and new export opportunities.* 2020 Brief 56. Washington, DC: International Food Policy Research Institute (IFPRI).

———. 2001. Rich and poor country perspectives on biotechnology. Paper presented at the preconference workshop, "Agricultural Biotechnology: Markets and Policies in an International Setting," at the Annual Conference of the Australian Agricultural and Resource Economics Society, January 22, Adelaide, Australia.

Pinstrup-Andersen, P., R. Pandya-Lorch, and M. W. Rosegrant. 1999. *World food prospects: Critical issues for the early twenty-first century.* Washington, DC: International Food Policy Research Institute (IFPRI).

Press, E., and J. Washburn. 2000. The kept university. *Atlantic Monthly,* March, www.theatlantic.com/issues/2000/03/press.htm.

Preston, T. 2000. An active eight years for IP. USPTO *Today Online* 1 (12): 3–7. Washington, DC: United States Patent and Trademark Office.

RAFI (Rural Advancement Foundation International). 2000. In search of higher ground: The intellectual property challenge to public agricultural research and human rights and 28 alternative initiatives. Occasional Papers Series 6 (1). Winnipeg, Canada: RAFI.

Ribaut, J.-M., and D. Poland, eds. 2000. Molecular approaches for the genetic improvement of cereals for stable production in water-limited environments. Mexico City: International Maize and Wheat Improvement Center (CIMMYT).

Rosegrant, M. W., M. A. Sombilla, R. V. Gerpacio, and C. Ringler. 1997. Global food markets and US exports in the twenty-first century. Paper presented at the Illinois World Food and Sustainable Agriculture Program Conference, Meeting the Demand for Food in the 21st Century: Challenges and Opportunities, May 28, University of Illinois, Urbana-Champaign.

Sandburg, B. 1999. The ivory tower IP fix. *IP: The Magazine of Law and Policy for High Technology*, May.

Savidan, Y., J. G. Carman, and T. Dresselhaus, eds. 2001. *The flowering of apomixis: From mechanisms to genetic engineering*. Mexico, D.F.: CIMMYT, IRD, and European Commission DG VI (FAIR).

Serageldin, I., and G. J. Perlsey. 2000. *Promethean science: Agricultural biotechnology, the environment, and the poor*. Washington, DC: Consultative Group on International Agricultural Research (CGIAR).

Serratos, J. A., M. C. Wilcox, and F. Castillo, eds. 1997. Gene flow among Mexican landraces, improved maize varieties, and teosinte. Mexico City: International Maize and Wheat Improvement Center (CIMMYT), Instituto Nacional de Investigaciones Forestales y Agropecuarias (INIFAP), and CNBA.

Smale, M., ed. 1998. *Farmers, gene banks, and crop breeding: Economic analyses of diversity in wheat, maize, and rice*. Dordrecht, the Netherlands: International Maize and Wheat Improvement Center (CIMMYT) and Kluwer Academic Publishers.

Smith, L., and L. Haddad. 2000. *Overcoming child malnutrition in developing countries: Past achievements and future choices*. Washington, DC: International Food Policy Research Institute (IFPRI).

Syngenta. 2000. Rice genome sequence fact sheet. January 26, 2001. Basel, Switzerland: Syngenta International AG.

Tripp, R. 2000. GMOs and NGOs: Biotechnology, the policy process, and the presentation of evidence. *ODI Natural Resource Perspectives* 60. London, UK: Overseas Development Institute (ODI).

U.S. Patent and Trademark Office (USPTO). 2001. Press release. January 19. Washington, DC: USPTO.

Winkelmann, D. L. 1994. A view of quintessential agricultural research. In *Technology policy: Policy issues for the international community*, ed. J. Anderson. Wallingford, UK CAB International.

NOTES

1. See www.cgiar.org and www.futureharvest.org.

2. For an idea of institutions with which CIMMYT collaborates, and for information on who funds our research, see our annual report (CIMMYT 2000), www.cimmyt.org/whatiscimmyt/AR99_2000/content.htm.

3. Morris and Ekasingh (forthcoming) point out that, unlike public breeding programs, private companies may not have placed much emphasis on breeding for

resistance to diseases and pests, especially if the companies included an agrochemicals division.

4. Chiefly wheat, rice, barley, maize, sorghum, millet, rye, and oats.

5. Paarlberg (2000), cited in RAFI (2000), reports that foreign aid to developing country agriculture fell by 57 percent between 1988 and 1996 (a drop from $9.24 billion to $4 billion, in 1990 dollars). He also reports that World Bank loans for agricultural and rural development fell 47 percent between 1986 and 1998 (from $6 billion to $3.2 billion, in 1996 dollars).

6. Demand for wheat will grow by 1.58 percent per year; demand for maize will grow by 2.35 percent per year.

7. It is hoped that international agreements such as the Convention on Biodiversity and the Food and Agriculture Organization's International Undertaking on Plant Genetic Resources for Food and Agriculture will contribute to their resolution.

8. The development of "golden rice," which contains higher levels of beta-carotene, a vitamin A precursor, was heralded as a major achievement on behalf of poor people. It took some time for the public to realize that poor people could not immediately gain access to golden rice because researchers had developed the rice using proprietary technology, and a host of licensing and other issues would have to be resolved before the nutritional promise of golden rice became a reality (see "New mechanisms for accessing and providing research products," later in this chapter).

9. See also C. E. Pray, 1999, "Role of the Private Sector in Linking the US Agricultural, Scientific, and Technological Community with the Global Scientific and Technological Community," unpublished paper, Rutgers University.

10. Corporations provided $850 million to U.S. universities in 1985 and $4.25 billion less than a decade later (Press and Washburn 2000).

11. An even more controversial arrangement had been established—and challenged—earlier between Scripps Research Institute and Sandoz (Sandburg 1999), in which Sandoz was first awarded the right to license all of Scripps' inventions over the course of ten years. This arrangement proved so controversial that, following U.S. congressional hearings, the terms of the agreement were altered to give Sandoz first rights to license 46 percent of Scripps' discoveries over five years, with an option to renew the agreement for another five years.

12. For example, Stanford University earned $61 million from its technology transfer efforts in 1999 (Press and Washburn 2000).

13. For a clear and thorough discussion of these questions, see Morris and Ekasingh (forthcoming).

14. Argentina, Brazil, China, India, Kenya, Mexico, South Africa, and Thailand, for example, have all dedicated significant resources to biotechnology research.

15. For example, a research consortium to address the challenging environmental and productivity problems in South Asia's rice-wheat systems has, at one time or another, involved national and international research organizations, nongovernmental organizations, farmers, local private machinery companies, and researchers from advanced public research organizations in industrialized countries. This consortium was recently recognized by the CGIAR as one of the most successful research partnerships established by the Future Harvest Centers. Information on the Rice-Wheat Consortium for the Indo-Gangetic Plains is available at www.rwc.cgiar.org.

16. Representatives of seven of the world's academies of science (the Brazilian, Chinese, Indian, Mexican, UK, U.S., and Third World academies) recently issued a white paper (Anonymous 2000) on transgenic plants and world agriculture, stating, "Private corporations and research institutions should make arrangements for GM technology, now held under strict patents and licensing agreements, with responsible scientists for use for hunger alleviation and to enhance food security in developing countries. In addition, special exemptions should be given to the world's poor farmers to protect them from inappropriate restrictions in propagating their crops."

17. In 1999, 3 percent of CIMMYT's research resources came from agreements with the private sector.

18. Details of these agreements may be found in CIMMYT's annual report (CIMMYT 2000).

19. For a technical overview of the quest to produce apomictic food crops, see Savidan, Carman, and Dresselhaus (2001).

20. See www.cimmyt.org/Resources/Obtaining_seed/IP_policy/htm/IPolicy_Eng .htm.

21. The authors are grateful to Victoria Henson-Apollonio, Senior Research Officer, Intellectual Property, with the CGIAR Central Advisory Service on Intellectual Property, for drawing our attention to patent pooling.

22. Perhaps this awareness builds on the belated realization by many pharmaceutical companies that they should develop more flexible and lenient policies for providing AIDS medication in developing countries.

23. See www.cimmyt.org/whatiscimmyt/tlaxcala.htm.

24. See, for example, Serratos, Wilcox, and Castillo (1997) on gene flow among improved maize, landraces, and teosinte; Dreher et al. (2000, 2001) and Ribaut and Poland (2000) on issues related to conventional and molecular breeding; and Smale (1998) on numerous genetic resources issues, including measurements of genetic diversity and the economics of genetic resource conservation.

7

Science and Sustainable Food Security

M. S. Swaminathan

Extreme poverty is the state of the general populace in India. Nevertheless, as serious as the poverty itself is, it is really the tip of the iceberg, a part of a complex system maintaining the economic and social status quo. The challenge for organizations like the M. S. Swaminathan Center thus has been to figure out which solutions would work within the Indian context, with all its particular constraints and conditions, not only to alleviate poverty but to equip people to survive. And to this end, the universities in India have played a pivotal role. Similarly, while the challenge of solving the many problems facing American society is given to the university, one of the goals of this chapter is that through sharing some of the changes that the Indian university has undergone, both can mutually benefit from each other's experiences and hone each other to become integral agents of change in society.

The top priority of an educator at an Indian university is teaching and training, that is, investing in the next generation of leaders. Scholars combine academic excellence with social and ethical concerns, as well as any other work that would scaffold these objectives. This type of value-based education is very important in today's technological world, because while there are always several very gifted scientists in every field at the cutting edge of the knowledge frontier for knowledge's sake, there are many more of us whose contributions are more practically applicable to the benefit of humanity.

Moreover, in Indian universities, outreach and partnerships have become exceedingly important. Universities can be instruments of changing public policy. And it is clear that there is a need for synergy between technology and public policy, because unless there is synergy between the two, there could be no technological payoff. UNESCO's 1999 World Congress of Science in Budapest called for a new social contract between science and society—that is, the development of a science for basic human needs. This is the heart of what we are trying to do through our centers, connecting the knowledge produced in these universities and making it accessible and applicable to the everyday village person.

It is no wonder that "sustainable science" is becoming somewhat of a buzzword in various circles of research. In fact, it has become so common to put the word "sustainable" before everything these days that we seldom pause and ask ourselves, what does sustainable mean? And looking back, the concept of sustainability has undergone considerable changes. While in the past it had an economic overtone, it later came to refer to "environmental sustainability,"[1] and now today it means "social sustainability," incorporating economic, social, and environmental components. Focusing on sustainable food security will hopefully give some insight into some of the challenges that we and our research universities face concerning social sustainability.

The concept of food security has undergone change. For a long time, food security meant having enough available food in the market. But today we know that having enough food in the market is not enough; we just have to look at the mass famines that have taken place in India, despite the fact that crop yields were very high. The problem is that people need money to be able to purchase the food. In fact, in many developing countries today, food security is best expressed in million person-years of livelihoods, and in million tons of food grains. In other words, buying power and economic access to food have become very important.

MEXICO'S WHEAT REVOLUTION

Before discussing the Indian situation in more depth, let us look at one example of how research can be done for the benefit of the public good. The wheat revolution in Mexico (which is discussed in chapter 6 by Dr. Timothy Reeves) has made great progress as a result of its partnership with the CIMMYT.[2] Dr. Norman Borlaug, Dr. Reeves' predecessor at the CIMMYT, was responsible for giving us the early semidwarf varieties of wheat, which in fact started the Green Revolution. Semidwarf wheats are wonderful grain types because, given enough fertilizer and water, they grow very proportionally and are able to convert most of their biomass into grains, called the harvest index. Most of the plant's energy as expended in photosynthesis then goes into making grain, producing high yields. Additionally, these varieties are also resistant to most pests and diseases. Because of the willingness to disseminate the agricultural R & D that was being done, this research contributed to the improvement of the quality of life in Mexico by ameliorating their agricultural resources and technologies.

INDIA'S GREEN REVOLUTION

India was about twenty-fifth in the world in regards to wheat production twenty-five years ago; today, it is second to China. And ever since 1968, when William Gaud of the United States coined the term the "Green Revolution," India's farmers have never looked back. We attribute this revolution to the rapid spread among small and medium-size farmers in India, since no large farms exist in India. One small farmer usually possesses two hectares or five acres or less. The marginal farmer then has one hectare or 2.5 acres or less. These marginal and small farmers constitute over 80 percent of India's farming community. And these are the farmers who have prospered year after year, for the simple reason that not only were their crops producing high yields, but their crops were also disease resistant. On top of these things, the government of India had a policy of minimum support price.

As I mentioned earlier, the concept of food security has gradually changed. Thirty or forty years ago, the UN's Food and Agricultural Organization (FAO) used to describe food security purely in terms of the surplus of grains available in the world. Today, the focus has shifted to looking at the function of purchasing power, where purchasing power is a function of income, employment, or livelihood opportunity. Even more recently, however, we have seen that despite having economic access to food and availability of food, without access to safe drinking water and environmental hygiene, the body can still be negatively affected. All three components, then—availability, access, and absorption—are necessary for food security.

We have tried to understand what the status of food security is in India. Currently, my colleagues and I are in the last stage of putting together an atlas detailing food and security. We have used nineteen different indicators, six or seven that relate to production, and the others to access. All of these facts are put together and considered in relation to each other. Almost a direct correlation emerges between a state's poverty and its degree of success in population stabilization. In turn, social expenditures and subsidies influence other aspects like clean drinking water and environmental hygiene. The highest level of food security is in Punjab, Haryana, and some of the other southern states. In our study, we have tried to examine food security from a two-dimensional perspective, looking at present yields and potential food security. The potential food security accounts for land degradation, salinization, and groundwater depletion, as well as groundwater pollution. Punjab and Haryana, which are located in the northwest region, currently possess high food security indexes, but they will sharply decline in the next ten to fifteen years if environmental sustainability is not addressed. If we do not attend to the environmental problems such as salinization and groundwater depletion, they will become major problems in the near future.

The "Green Revolution" is a term that means many things to different people, but essentially, in layman's terms, the "Green Revolution" refers to productivity improvement, not through area expansion, but through vertical growth in productivity. These components are relevant in developing countries because a very large percentage of land is already dedicated to farming, and there is no extra land that can be reallocated. Rainforests have dwindled and continue to dwindle. Therefore, there is a need for increasing production yields through productivity improvement.

The Green Revolution was, in fact, commodity centered—that is, on wheat, rice, corn, sorghum, and so on. However, a new revolution is now needed, a revolution that continuously increases productivity without any associated ecological harm, and where production technology and foundations can be developed in an integrated natural resources management system. The land, water, flora, fauna, and atmosphere are all very important and crucial. I would say that this revolution could be called the "Ever Green Revolution"—in other words, a productivity revolution that would be self-sustaining, leading to the formation of a true ecosystem.

India's strategy toward an Ever Green Revolution has three basic steps. The first step is to defend the gains already achieved. Vigilance is the price that needs to be paid for good agriculture. You may have a new pest, disease, or soil problem like salinization, which thus requires "maintenance research," or, in essence, defending any gain. The second step is to extend these gains to new areas, particularly rain-fed areas, other hilly areas, islands, and so on. And the third step is to make new gains, particularly in terms of income and employment through intensification, diversification, and value addition to primary products in farming systems.

THE GREAT DIVIDES

The Demographic Divide

Social instability can then be divided into five divides or categories because sustainability has to be defined on economic, ecological, and social terms. The first divide is the demographic divide. Globally, there is a demographic divide between industrialized and developing countries; in India, there are even demographic divides among the states within the country. For example, one of the greatest challenges in relation to this demographic divide is attracting and retaining youth in farming, because they are not interested in farming anymore. It will be very difficult to get educated youth to go back to farming because they feel that because they are now educated, farming is neither economically rewarding nor intellectually stimulating. For this reason, modern ecological farming has become very knowledge intensive, and in fact knowledge has become a substitute for chemicals in many cases. This is a more attractive method of getting youth back into farming. The other challenge is the greater need to pay attention to urban, paraurban, and semiurban agriculture, and since 50 percent of India's population will be in urban areas, there is great potential here. Therefore, one must work the demographic trends in relation to our own research priorities.

The Economic Divide

The economic divide is the most difficult one in the world today because the rich-poor divide has grown quite large. Statistics show that 20 percent of the world's population has more than 80 percent of the world's income, and another 20 percent has less than one percent of the world's income, which is what the World Bank calls less than one dollar per capita per day, forming the United Nations Development Program (UNDP) champagne glass.

The late Dr. Marbel Huck, when heading the UNDP Human Development Report committee, pointed out that the world was witnessing what he referred to as "job-less economic growth," where there is economic growth, but the number of jobs is not increasing. Even some industrialized countries are having problems with joblessness or high unemployment rates. What countries like India, China, or Bangladesh need is a "job-led economic growth," not jobless growth.

The Gender Divide

The third divide then is the gender divide. Here again, it is important to note that if we really want to address the needs of the poor in India and many other developing countries through scientific research, it is important to have a pronature, propoor, and prowoman orientation in technology, development, and knowledge dissemination. The Beijing Conference on Women (1995) and the Beijing Plus 5 Conference (June 2000) in New York clearly reflect the trend toward feminization of poverty and agriculture. Gender studies need to be done in our research agenda because there is a gender divide in many towns in India where crops are cultivated resulting from certain cultural norms; that is, the food crops are taken care of by women and the cash crops by men. We have also asked universities in India to incorporate gender studies into their agricultural curriculum at the bachelor's degree level and above, because unless it is incorporated into the curriculum, the importance of this gender dimension in agricultural advances may be overlooked.

The Genetic Divide

The fourth divide is the genetic divide, represented by a schism between modern genetic information and the dissemination of that information so that it can become public knowledge. For new varieties, hybrids, or other types of biopesticides, biofertilizers are important biological technologies that can be traced back to work in genetics. In one area in India, we have developed three different approaches to bridging the gap between modern genetic information and disseminating that information to the public, starting in rural areas, in what we call a "biovillage." "Bios" means "living," and a biovillage, then, is a "human-centered development, looking at both natural resources conservation and poverty eradication as two sides of the same coin."

The residents in these biovillages become the decision makers by trying to bridge the genetic gap by connecting research, disseminating that knowledge, and then applying it to produce resources. Their felt needs are ascertained, determining their role in rural service. The activities they take up are market driven so that they are not just based on subsidies, and the beneficiary approach of development based on patronage then gives way to an approach regarding rural inhabitants as producers, innovators, and entrepreneurs. This is exceedingly important. This type of change in mindset is important because often administrators and political leaders call the poor "beneficiaries" or "target groups." But it is important to recognize them as innovators, whether they are literate, semiliterate, or illiterate, and that people take to new technologies like "fish to water," provided the pedagogic methodology has a hands-on learning approach and does not take place simply via lecturing. We currently have a whole series of biovillages in certain parts of southern India.

Dr. Reeves rightly mentions that 70 percent of India and many other developing countries depend on agriculture for their livelihood. Thus, so long as they are operating via their status quo farming operations, they will remain poor. But by incorporating new technology and knowledge, the agricultural yield for the entire village could be improved. For example, in some of the biovillages, farmers have been discouraged from growing rice and sugar cane because these are both water-loving plants. What consequently happens if farmers do not desist is that the water supply, a precious commodity for most farmers in developing countries, becomes depleted for the entire community. And so they have been encouraged to cultivate "high-value plants," such as legumes, which require little water.

The Ever Green Revolution will depend on incorporating and integrating different approaches. Thus, crop diversification and value addition have become important for shifting some of the farm people away from the routine operations of farming. One example of this practice can be observed in the development of derelict ponds in biovillages into aqua-culture ponds, separated by layers and different techniques, allowing the water to be used in the most effective way. The results are processed by the villagers, who then take the goods to the urban areas as a means of supplementing their incomes.

Mushroom production is another example of value addition, and this has become very common in these villages, particularly with extremely poor people. All that many of these people possess are their huts and these paddy straw mushrooms. Mushroom production gives them another thirty dollars of additional income per month, which is very significant in India for such people.

Within these local biovillages, then, there is a single-stop service center, which is called the biocenter. A biocenter is where the people themselves provide the training and materials needed for producing these different types of value additions. They provide the seed for the mushrooms, stock for earthworms if there is a worming culture, all the hybrid seeds, and even artificial insemination for cattle. Within these biovillages, then, everyone becomes empowered to work and contribute. The biocenter is, in essence, the hub of the biovillage movement because, again, it is operated by the people themselves, and we have found in the last six years that once you develop a strategic research program, it is better to turn the program over to the people whom it will be affecting as soon as it is feasible.

The goal, then, is to scaffold the program for a few years until it becomes self-operating. Most externally funded and supported projects collapse when the external input is withdrawn. In our research agenda, then, it has been essential to find out how to keep these projects sustainable over a long period of time, regardless of whether the original people who started it are there or not.

These bioparks are similar to the industrial parks in the United States. The first of the bioparks we helped establish was an all-woman biotechnology park, about twenty miles from the city of Madras. We also have three major biovalleys. One of them is being developed with support from Cardinal University and is the University's own biopark. This is also near Madras and largely functions as a genomic center, working on bioinformatics and more sophisticated biotechnology. Of the other two, one is a marine biovalley, producing whole series of marine products like eel, prawn, crab cultures, and the like. And the third one is a medicinal plant biovalley. The medicinal plant biovalley is particularly important because the southern states of Kerala and Tamil Nadu, as well as sections of the eastern and western parts, are very rich with plants with medicinal properties. These valleys provide infrastructure for using these medicinal plants, including procedures for validation. Validation of claims of these plants is very important in order to ensure quality control management.

However, there is a growing tension concerning intellectual property rights in relation to these bioresources, and one particular area this is becoming more prominent concerns the aforementioned plants. It is important to recognize and consider whether injecting the word "medicinal" before another word should trigger rights and concern. In nature there is only type of plant, not medicinal plants or other types of plants. Medicinal value arises from traditional wisdom, knowledge, or indigenous knowledge.

For example, the *trachotis* is a plant that one could use to remove fatigue. In essence, this can be used as an antifatigue plant. To improve memory, one of our famous chess champions also uses this plant as a memory enhancer. The injustice lies in the fact that the plant gatherer gets a very small fraction of the total money because the merchant and exporter are the ones who get most of the money. This is where ethics and equity are incorporated into the equation, because in this case (and oftentimes) the primary conservers of the genetic materials remain poor, while those who use their knowledge and material, either through normal Mendelian methods or biotechnological methods, become rich. It is this type of unethical situation that the Convention of Biological Diversity has tried to alleviate for the first time by adding amendments on ethics, equity, and benefit sharing.

Finally, in addition to using technology and knowledge to bridge the genetic divide, it is also important that there is a diversity of genetic material available. Monoculture and genetic homogeneity is an invitation to genetic vulnerability to pests and diseases. Diversity is therefore very important, and this is best achieved through participatory breeding programs with farming communities. Institutions like the Cardinal University, along with its Biotechnology Department, can prepare novel genetic combinations through transgenic mechanism. Using that novel genetic mechanism produces a large number of locally adapted varieties jointly with farmers, and that way you can combine both genetic diversity and genetic efficiency.

Let me again summarize by saying that the world is getting more and more divided, and, especially economically, the world is becoming more stratified. Again, the UNDP champagne glass illustration showing the breakdown of the world's population and distribution of wealth is the most pictorial assessment showing that there is no noticeable improvement. Even if there is an improvement in percentage terms, values in percentages are oftentimes misleading. In India, one might say, "Hunger has been reduced by 5 percent." Unfortunately, lower relative numbers like this for something like hunger or literacy are simply indicative of larger absolute numbers because of the overall population growth. Therefore it is important to discuss numbers rather than strict percentages.

Mahatma Gandhi had a very good recipe concerning people's development into scientific technologists. Recall the face of the poorest, weakest man that comes to mind, and ask yourself the following questions: Are the steps you are contemplating going to be of any use to him? Will he gain anything by it? Will it restore the control of his life and give destiny back to him? In Mahatma Gandhi's case, he not only posed the question, he offered a solution in providing a simple concept that any scientist could practice. Our mission statement is "to impart the propoor, pronature, prowoman orientation to technology development and dissemination." Therefore, before we start any work, we evaluate: Is this pronature, propoor, and/or prowoman? Once you have the most important criteria satisfied, everything else then falls into place and everybody benefits. Research universities might consider the same motto.

The Digital Divide

The fifth and last divide I will discuss is the digital divide, and similar to the genetic divide, this refers to the gap between existing digital technology and knowledge accessibility for people. And so we developed a cyberpathway in order to attempt bridging the digital divide. We went to villages, talked to the poorest people there, established hubs like the biocenters, explained what these hubs were and how to use the Internet along with the various services, and empowered the village people to operate these things. They mastered these skills within thirteen to fifteen days. One such village was a fishing village. A woman told us that because all the information in their village is demand driven and need based, they themselves decide what information they want, and in this case, they wanted to know how to access information about the sea's condition. The men go out in catamarans and are not mechanized boat fishers. The women were saying that because of this they did not know whether the men would return by the end of the day, particularly during the rainy season. We showed them how to access the pertinent information, and since that information is fairly accurate, they have used it for the last two years, downloading it from the Internet every evening. This information is then broadcast all over the village through the loudspeaker, describing the sea at one hundred meters from the coast, one thousand meters from the coast, and so on. New technology has practical and tangible relevance to their lives.

And all these things can be used for educational purposes. I think that bridging the digital divide has tremendous implications for bridging the demographic and gender divides, at least under our conditions in India, since there is a lot of social discrimination, yet through these things, women can be empowered to become knowledge providers.

CONCLUDING THOUGHTS

In conclusion, whether we are talking about genomics, proteomics, the Internet, or nanotechnology, we are entering into an absolutely fascinating era of science. It is amazing to see the exponential growth in all these areas, and all of them can provide an unprecedented number of opportunities for a sustainable human future for humanity if strong ethical pushes are applied to technological pulls. This is the most important requirement for the changing priorities of universities and their changing value systems. The university not only should be on the forefront of a technological pull, but also it is very important to have an ethical push, so that one can arrive at one's technological goals without compromising the public good. Indeed it is a fascinating world of science we have entered. I do hope that in a world of increasing intellectual property rights and protection, exclusive knowledge, that this kind of partnership among institutions committed to the public good will not only continue but will be furthered to the benefit of mankind.

Following are Dr. Timothy Reeves' (TR) and Dr. M. S. Swaminathan's (MSS) responses to questions posed during a discussion of their papers at the University of Illinois Center for Advanced Study.

Q: What exactly is/was the Green Revolution?

A (MSS): *The term "Green Revolution" was coined by Dr. William Gaud, a scientist from the U.S. Department of Agriculture in 1968. The very word "revolution" indicates it was not an evolutionary change. There was a certain quantum jump in production productivity. In India, for example, production from 1965 to 1966 was about 10 million tons, and then 17 million tons in 1968. Since this was not a small incremental change, it was considered a revolutionary change. Green is the color of chlorophyll, and the green plant, which absorbs sunlight from the atmosphere, takes water. The reason for this change was a new kind of plant architecture, the semidwarf structure. This is because the earlier varieties were all tall, and so when they were placed in fertile soil and watered, they would fall down. This is technically called lodging. Thanks to Dr. Norman Borlaug and Orville Vogel in Washington State's Pullman, there are new semidwarf wheat varieties. Dr. Vogel developed winter wheats, and Dr. Borlaug spring wheats. Dr. Borlaug also introduced a photo-insensitive character, for which he was later awarded a Nobel Peace Prize Laureate in 1970. In other words, this variety can grow irrespective of the number of hours of sunlight. This made a significant difference in India and Pakistan, and at the same time there was also a rice revolution in progress; thus, it was called the Green Revolution. In India we called it the Wheat Revolution first, and then later on the Rice Revolution. But the Green Revolution is the generic term indicating the fairly rapid progress in improving production, but through productivity improvement or yield rather than by area expansion.*

Q: I guess that leads us to the question some people have asked, "Are we near the point where we have increased production as much as we can? How much more can it increase?"

A (TR): *That is a very interesting question, and there are two aspects: on the one hand, we can approach this question in terms of yield potential, and the production of new types of plants with tremendous capacity to give greater yields and quite often with lesser inputs; but the other part of it, of course, is making that happen in farmers' fields, and quite often these are two different issues. So I would say that in rela-*

tion to both, we certainly have not reached the peaks yet. New technologies are giving us different crop varieties and technologies that can significantly increase yield potential. Dr. Swaminathan talked about new plant types, and we are currently looking at new plant types again now because we believe these particular ones could give up to 30 percent increases in yield. But one of the major challenges around the world is to actually get things happening in farmers' fields. And one of the things we see is that the investments in agriculture, that is, the investments in the processes to develop new technology and get farmers to use them, are, in fact, falling dramatically around the world.

Q: Let me ask one of the criticisms that has been raised about the Green Revolution. Essentially what has happened is now that farmers around the world, particularly farmers in developing countries, have become increasingly dependent on technologies coming from the developed world, that is, inputs with fertilizers and chemicals with highly advanced plant varieties, and that that is a problem. Is that a fair criticism, and is it in fact a problem?

A (MSS): *It is neither a problem nor a fair criticism for this simple reason: the plant is not a magician, and inputs are needed for outputs, whether it is a developed or developing countries. Some of the developing countries, like India, have certain advantages being in the sans and sub tropics, which have sunlight throughout the year. This means that we can grow several crops a year, unlike in the temperate regions. So you need inputs. For example, fertilizer applications started around 1870 in Germany, but there was a bad gap in the application of these technologies in developing countries. But pressure came as a result of population growth. After World War II, the population started rising very fast because while the mortality rate went down, the birth rate did not, and therefore there was a need to produce more from the same land, water, et cetera. In other words, exactly what Dr. Reeves was talking about when he discussed productivity improvement. If you take all the additional land, any remaining forest would have disappeared. This is why we call the Green Revolution a forest-saving form of agriculture. For example, for India to produce 75 million tons of wheat as we produce now, at the yield rate of one ton per hectare, which was common before the Green Revolution, 75 million hectares would be needed, but the present 75 million tons is being produced by 23 to 24 million hectares. Therefore, as the population pressure on land increases, we have no option except to produce more from diminishing land and per capita water resources, and this has to be done through nutrients, which then need to be applied. Biological control, genetic control, has been very important. In fact, the reason why India had the Wheat Revolution rapidly was because the varieties were originally obtained by the CIMMYT in Mexico, and were also bred within the country. As a result, they had a broad spectrum of resistance to all the major pathogens, which is why farmers took to it even faster because they saw the disease resistance of these new varieties. So I would say, in technological terms, we cannot make a division between north or south, because a plant is the same physiological entity—it requires certain nutrition, it requires some sunlight, it requires water from the soil—and developing countries were just a little later in the arriving at this juncture in terms of technological transformation of agriculture.*

A (TR): *Picking up on one of Dr. Swaminathan's comments, which is really important, is that at the farm level you can have a sustainable system, if it has high inputs and outputs, or if it is low inputs and outputs, but you cannot have sustainability on these farms with low inputs and high outputs; and yet these farmers have to increase their outputs so that you can balance within that system. So what we are doing now is what Dr. Swaminathan was just saying, creating plants that are resistant to disease so that fungicides or pesticides are not needed, more efficiently utilizing water and turning it into yield, more efficiently using fertilizers and turning it into yield, better tolerating heat and drought, et cetera, all increasing the reliability of production, guaranteeing that it is increasing as well as actually the productivity itself. And these are major emphases of the work as it goes on.*

Q: Something you touched on a couple of times that I know is a subject that concerns a lot of people around the world is regarding water. How much water is used by agriculture, and what are your thoughts on the idea that increased production will require more water?

A (MSS): *Water is definitely going to be a key constraint in the future in many parts of the world. Even for domestic consumption, water is becoming scarce. Agriculture normally consumes nearly 80 percent or even more in many countries as far as water use is concerned. There is a greater awareness today of the need to use water in the most effective way. Countries like Israel were pioneers in getting more crop per drop, as they would say: maximizing every drop of water, through very efficient use of water, along with, now, genetic improvement of their grain resulting in grain production per unit of water. So you have to have a multipronged strategy: First, conserve and save every drop of water. Second, use it in the most economical way; for example, where it is drip irrigation, there tend to be greater advances in terms of using water in an efficient and effective way. Third, make the plant more drought tolerant so that water is used more efficiently and the cropping system is designed in such a way that you can maximize income and yield per every drop of water available. This type of consciousness is spreading. It is slow in some cases, and there is still much of waste. For example, in my own country, unfortunately we say that it pours, and it never rains. This is very highly skewed, since all the water comes during a monsoon—basically, the whole year's rainfall in a few days. So unless you conserve and preserve the water, you will be in a bind. So there are watershed development and water conservation movements forming. I am hoping that the scarcity itself will trigger new innovations and practices of economic water use.*

Q: I think one thing that is certainly on a lot of people's minds is the fact that a lot of the research on these plants is being done either by private companies or by partnerships between universities and private industry, and that the profit motive is very strong. If it is really all about money, then will we be able to develop the kind of plants that we need to feed the world, and can we find the particular varieties suited for particular places?

A (MSS): *In India, we are convinced that unless there is concurrent investment in what we call the public goods research, like what the CIMMYT under the leadership of Dr. Reeves is doing, and a national agriculture research system, then some of the important problems of these resource-poor farmers may not get attention. Because you are quite right: if the market is the major determinant of investment decisions, then orphans will remain orphans. That is why there is need for a balance between public-sector and privately sponsored research. But there can be synergy between the two. But this is an area where, particularly in countries with very large numbers of resource-poor farmers and a number of location-specific problems, publicly funded institutions like universities should be adequately funded and priorities set according to the problems troubling the resource-poor people. If that does not happen, as in the early days of the Green Revolution, social scientists criticized it by saying the new technology has made the rich the richer and the poor the poorer because that gap between the rich and the poor has become more stratified. We should avoid any such thing happening with the new biotechnology, or the gene revolution as it is called, especially with the genome maps being unraveled, because there are enormous opportunities here to address some of the critical issues facing resource-poor farmers, that is, through the creation of drought tolerance, salinity tolerance, disease resistance, et cetera, as properties of plants. So I would say that, like everything else in life, there has to be a balance between private-sector research, motivated by adequate income or profit, and the public-sector research that addresses issues which otherwise may not get attention.*

A (TR): *Balance is really important. There is a tendency around the world for, as I said earlier, a public investment in agriculture, generally in agricultural research particularly, to decline. There is sort of a mentality across the world amongst decision makers and politicians that food comes from supermarkets, you know, and it will always be there. And so, given that scenario and the complexity of the problems we deal with, it is absolutely vital to have partnerships, but in a way that levers the resources of the industrialized nations onto the problems and opportunities of the developing nations, and how we can do that? Because the same technology that may help a company capture the market in North America could probably be used to find some other trait that would be beneficial for poverty-stricken farmers in Africa, and that is the challenge for us.*

Q: Something that is difficult to understand but that is inherent in the idea of the Green Revolution in developing particular things and then broadly applying them is that, as that process proceeds, farmers over wide areas will replace thousands of traditional varieties with a few of these newly developed varieties. It seems that two kinds of risks are then introduced. The first is that an unforeseeable pest or a disease can arise, and the second is that this can cause huge losses in a short time, so that when the researchers try to address the problem by developing something that can resist the new attacks, the genetic variety that previously existed is now gone because nobody has sustained all the thousands of varieties that were replaced. How do you look upon this problem?

A (MSS): *I think that is a very important question, and first of all, let me say, from our experience in India, the term "Green Revolution" was coined in 1968, so we have thirty-two years of experience. Whether wheat or rice, productivity and production has gone up from 10 to 75 million tons, and in the case of wheat, from about 25 to 80 to 90 million tons now. And the reason why that type of situation has been avoided is partly because the new varieties have a very broad spectrum of resistance; if you see the pedigree, it is very complicated. Secondly, there are a large number of varieties even among the high-yielding varieties. It was not like this in 1968; there were maybe three or four; now several hundreds are there. And thirdly, the land races have all been collected and conserved in gene banks. For example, institutes like the CIMMYT have a whole family of the consultative group that has nearly 600,000 accessions; the gene bank at the International Rice Research Institute has over 100,000 varieties of rice stored in them. So, all these species are preserved somewhere. And essentially many cultivators in marginal environments and in more high-risk areas will continue to cultivate local strains that have adjusted to fragile environments, and this has been deliberately promoted. So while it is true that the number of varieties cultivated from the past is decreasing, these land races are being conserved, and secondly, the pedigrees of new varieties have become very broad based, bringing a large number of earlier cultivars into the genetic makeup of the new variety. This is probably the reason why we have not seen any regression. Production has been increasing every year, not negatively affected by any type of disease problem. There have been soil exhaustion problems, but no major disease outbreak has happened.*

Q: Professor Reeves, you said that there is a need to persuade governments to put more public money into agricultural research. How exactly does one do that, given the fact that it is going the other direction?

A (TR): *That is a difficult issue. The network of voters, of governments, et cetera that extends to the rural communities is remote, at best, in most countries of the developed world. And although you still have 70 percent of the population dependent on agriculture for their living, their political clout is generally quite low. And so, in many of these countries, the governments believe that they can leapfrog from subsistence smallholder agriculture, tertiary industry, and then to information technology, et cetera, but it does not happen that way. For India, agriculture is the engine room of economic growth, and the economy can take off from there. So if we implement these changes, we would convince people that it is not just about food, but about helping to alleviate world poverty. And because over 70 percent of people in these countries depend on agriculture, it is not just food supply; it is their earned income. And if you are going to improve their living and help them have disposable income available for health, for education, et cetera, the most powerful tool then becomes improving the agriculture. This requires both private investment in places where the market can assist that and public investment in places where the market will fail.*

A (MSS): *In India, the media play a very important role here. They warn the political leaders and administrators that they cannot abdicate their responsibility to step up public good research using new technologies, whether it is genomics, proteomics, or Internet and nanotechnology all the time. The media also certainly play a very critical role in democratic society in ensuring that the voice of the public is heard by leadership. It is making a difference, and there is a growing realization that there must be a balance, as I mentioned earlier, between research for the public good and research motivated by commercial opportunities.*

NOTES

1. This term was used particularly after the Stockholm Conference on the Human Environment and, more recently, the UN Conference on Environment and Development.

2. CIMMYT is the International Maize and Wheat Improvement Center located in Mexico City, Mexico.

Section III

FUNDING, ECONOMIC INCENTIVES, AND THE RESEARCH AGENDA

Kathie Olsen and Ann B. Carlson open this section with an insider assessment of federal science and technology funding policy, a major driver of university research agendas. James Savage continues with a powerful indictment of federal earmarking and its effect on university research agendas, combined with an eloquent plea for the restoration and strengthening of peer review. Rebecca Eisenberg closes with a comparative look at how the different incentives affecting university and private research pursuits interplay at times to promote the free public dissemination of private-sector discoveries and the intellectual property protection of research university discoveries.

8

Federal Science Policy and University Research Agendas

Kathie L. Olsen and Ann B. Carlson

If we look at the big picture of the role of the federal government in science and technology, it ultimately boils down to improving the nation's ability to innovate. To perpetuate innovation, it is essential for the federal government to support high-risk research activities. And in order for this support to be effective, there must be accountability: we must be able to monitor what we do, analyze it, and see over the horizon. There are additional roles for federal research in ensuring national security, ensuring the health and well-being of our citizens, developing an educated society, and enhancing the economy through the development of enabling technologies that then can have numerous applications when transferred to industry or other sectors. In all this, we must maximize investments and ensure cooperation across federal agencies, and with industry and academia.

Accountability is critical. A major new law that has significantly impacted federal research agencies is the Government Performance and Results Act (GPRA). For government agencies, the GPRA has become the yardstick measuring what we do and how well we do it. The aim is to develop measures for the success of our programs and activities, and then to report the outcomes annually. It poses a unique challenge for the science agencies, since science as a discipline has never worked on a timetable, nor can major scientific findings and discoveries be predicted in advance. Although measuring unknowns is always difficult, we must try, and we are learning how to accomplish this task every year. Congress understands and is sympathetic toward this dilemma that the science agencies face, but nevertheless, they still want today's activities to produce tomorrow's product. It is American tax dollars providing the federal investment in research and development (R & D). Thus, it is important for Congress to understand not only what they are getting for their investment, but also what they are getting "today" and not just "someday." This is the reality of the world that Congress lives in. The federal science agencies must, in turn, be responsive. Thus, the GPRA is taken very seriously. We are always looking for advice to improve our ability to measure performance, and indeed, we are taxing our agency's advisory committees for guidance. We are struggling with how best to demonstrate scientific output and return.

The White House Office of Science and Technology Policy (OSTP) and the Office of Management and Budget (OMB) are essential in planning and coordinating the full range of the federal science and technology (S & T) investments. Setting priorities across scientific disciplines, or even within disciplines, can be very challenging. There are those who consider it practically impossible. OSTP and OMB, each year, provide coordinated guidance on the direction of S & T investments among and between the federal agencies and help guide the priority setting needed in the federal government.

UNDERSTANDING THE PLAYERS

Amidst setting science priorities, it is also important to realize that the major funding decisions of the federal government will always be public, and they will always be political. After all, politics is simply the art or science of governing. Constituents represent their interests and priorities to their legislators and to the agencies funded by their taxes, while the government tries to act in the best interests of all. We need to take a look at the roles of the different players in this process, including the role of the university. Universities have a major role in both the public and the political process.

Figure 8.1 shows some data from the NSF comparing the U.S. investment in S & T to other countries. As you can see, the U.S. investment is substantial, and the rate of investment is increasing. From the results of this investment, we have obtained the status of a world leader in S & T.

As you can see in figure 8.2, we have a history of emphasizing different fields in different periods of time. Back in the 1960s, most of the money going into R & D was allocated to the space agency. The Russians had launched the first satellite, and we were going to "beat" them. Thus, we were going to be the first on the moon. Space was a major S & T priority. If you remember standing in line every other day to fill your car with gas, you understand why from 1973 to 1979 our priority was investment in energy research. During this time period, the Department of Energy (DOE) was established, and energy was the new priority for federal R & D investment. From the late 1970s until the end of the cold war, it was defense that dominated federal science investment, while the biological and biomedical sciences were losing students who were considering research careers because of limited funding

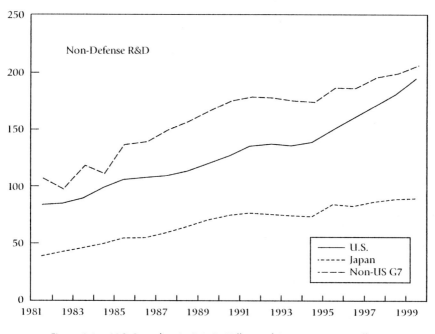

*Figure 3.1. U.S. Spending in R & D (Billions of Constant 1995 Dollars)
vs. That of Other Countries*

opportunities. The success rates on NIH grants were about 20 percent and were even lower for some NSF programs. More recently, the major increases in federal funding have been primarily at the NIH in the areas of health, bio-medical, and biological sciences, because advances in genomics and tech-nologies have set the stage for future breakthroughs.

I was surprised, once I did the research, to discover the substantial level of U.S. R & D investment by industry. Statistics provided by the NSF indicate that private investment accounts for 89 percent of the increase in all U.S. R & D from 1993 to 1999, but mostly in applied research and development. In contrast to the private sector, a major role of the federal government is to provide support for basic research, what we like to call "discovery research," fueling the engine of U.S. innovation far into the future.

Federal funding for R & D comes out of a very small pot of discretionary funds relative to the entire federal budget. A typical federal budget might provide a 2 to 5 percent increase in discretionary R & D funds. The amount of the increase is then divided among all federal science agencies, with NIH, NSF, and DOD receiving the lion's share. This "divvying up" is stressful for the federal agencies, especially when one considers how it goes through Congress. There are thirteen appropriations bills. NIH is part of Health and

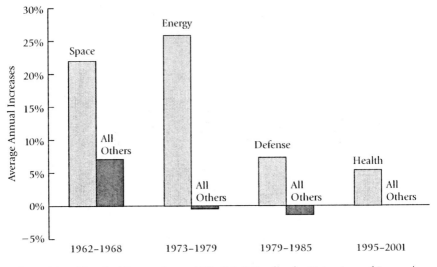

*Figure 8.2.   Historical Perspective on U.S.: R & D Funding by Major Area of Research;
R & D Balance Includes Setting Priorities*

Human Services (HHS), and they are competing with a pool of money that is appropriated for labor, HHS, and education. There is a set budget for money allocated for these areas, and the relevant appropriations committee then divides this money accordingly. On the other hand, NSF, NASA, EPA, HUD, VA, and all the other independent federal agencies compete against each other for the defined amount of funds available to their appropriations subcommittee. It is a very difficult job, given the decisions that must be made to allocate funding between the aforementioned agencies. The DOD R & D budget is under the umbrella of the full defense budget, and although it gets "decimal dust" compared to the total, it still does pretty well, especially within the development category. DOE is under Energy and Water Development, which defines the parameters for how their budget trades are made.

The time between the president's initial budget proposal and Congress's disposition of the budget can be quite exciting. Congress has routinely shown relatively strong support for federal R & D, but one never knows quite how the pie will end up being sliced.

FUNDING VERSUS EARMARKING

In the last several years, there has been an incredible increase in congressionally directed projects, or "earmarks." For federal agencies, earmarks feel something like having to pay taxes on capital gains when you have lost money on the stocks—they come directly out of the agencies' allocated budgets, often making budget "increases" illusory.

This is a challenge for universities as well. Universities lobby for earmarks, while their faculty members also compete for federal research grants. Unless Congress allocates new funds to support the directed research, it will become a balancing act or a zero-sum game. Universities play an important role in educating the public with respect to the importance of research and its integration into education. Universities also play an important role in developing the bottom-up approach and maintaining the integrity of the peer-review process in establishing research programs and opportunities.

While addressing congressional staff, some scientists tend to advocate an increase in support of program X, even if it is at the expense of support for some program Y. Scientists have the reputation for constantly demanding more support for their own programs. In fact, I have heard it said that there is one thing upon which all scientists agree: that more is better. But this, in the long run, hurts research. Divided, everybody loses. Instead, we need a balanced portfolio that includes the whole range of fields (e.g., engineering, physics, biology, chemistry, medicine). We need to capitalize on the breakthroughs in biomedical research as reflected in NIH increases while still maintaining a balanced research plan to ensure a vital research program that will prepare us for tomorrow and stimulate our economy. We oftentimes say that science is driven by technology. But in reality they go hand in hand. Physicists, chemists, and engineers all provide us with the basic knowledge and technologies that allow new developments in other fields, such as medicine. Without investment in basic research, new discoveries and cures will not occur as readily. We all need to watch emerging trends in engineering, physics, chemistry, life sciences, environmental sciences, mathematics, and computer sciences and advocate for each other by presenting a united agenda for science and technology to the Congress and the public.

The federal R & D budget is broken down into four categories: basic research, applied research, development, and facilities and equipment. One thing that is clear is that there is not universal agreement on the definitions, particularly for basic versus applied research. Depending on the administration, there may be more emphasis on basic research or more emphasis on applied research. How the agencies report basic and applied research can make a big difference in how well a program fares in the appropriations process.

At NASA, there is some tension over the issue of how research is reported, because OMB thinks of NASA as a mission agency that is mostly applied, while many NASA researchers believe that what they do is primarily basic research. For example, NASA believes that most of its climate change research is basic, because there is still so much that we do not know, and so much left to discover. However, the research is carried out to support NASA's mission to help decision makers with new, more accurate information. Thus, the outcome of the research within NASA has an applied aspect. As a result, NASA has proposed a new term, "targeted basic," which is basic research targeted toward a mission. NASA developed this idea because it believes the change would make its reporting more understandable. However, for a lot of other reasons, OMB may not be ready for another reporting category. What is essential is that there is understanding between the various parties about both the purpose and the level of maturity of the science, so that expectations for results are realistic and the programs can be properly evaluated.

Finally, the "R & D facilities and equipment" portion has been cut in recent years. At NSF, the cuts have been quite dramatic in the budget that brings state-of-the-art multiuser equipment to universities. Although universities need equipment and facilities, there are very limited opportunities within the federal government to apply for funds to support such projects. As a result, the federal agencies are beginning to see earmarks for these areas. If facilities and equipment are critical for R & D, it may be time to begin developing federal programs that can address these challenges, ones that can be introduced by the agencies to be included in the president's budget.

UNDERSTANDING AGENCY PLANNING AND PRIORITIES

The next question, then, is how are federal science agency allocations made? Federal government agencies have two basic approaches: top down and bottom up. The bottom-up approach uses inputs from academia, industry, and from within the ranks of the different agencies. Promising ideas are identified and passed through a preliminary development stage, maybe a concept study, and then the scientists and engineers try to get the idea passed upward through various levels of management and then incorporated into the budget. At the same time, we get the top-down approach from the OMB, OSTP, and Congress. They help set our priorities. Each agency's process has aspects of both the top-down and the bottom-up approaches, but they operate differently. The NSF and NIH favor the bottom-up approach. If they hear that an area is ripe for investment via advisory groups and workshops, especially those that are interdisciplinary (e.g., bio- and nanotechnology or information systems), they will take those priorities back to their strategic plans and incorporate those ideas into the budget process.

With the DOD, DOE, and NASA, the top-down approach is emphasized more strongly. For example, it is predetermined that NASA needs a second-generation launch vehicle, so the agency invests in the technology that will take us there. But even so, with scientists you cannot stop ideas from filtering upward, and at NASA, the bottom-up approach is also encouraged.

NASA's decision priorities include science return, benefit to society, alignment with NASA's mission, partnership opportunities, outreach, technology readiness, program balance, cost, budget, and of course mandated programs. In implementing these priorities, I noticed two distinct cultures at NASA. To me, as a scientist in an engineering-dominated environment, these cultures appeared quite distinctive. About $4 billion of NASA's $14 billion budget is for scientific research. Scientists care primarily about quality, so about 90 percent of NASA's science budget is peer reviewed and under open competition. In caring about quality, scientists have developed an attitude that says, "Quality has no fear of time." Scientists drive engineers crazy. Scientists always want to continue improving things by changing hypotheses and by "winging things." This is in contrast to engineers, who adhere to the concept that things are "essentially done." For example, the space station is essentially done. There may be limited research facilities up there, but to the engineers, it is essentially done. And then there are risk-assessment and performance factors, which are important to engineers and to NASA but aren't part of the scientist's vocabulary. Yet, despite their differences, the two camps work together in synergy on important projects that have never been done before.

NASA works very closely with other agencies. The Biological and Physical Research Program alone has over twenty memoranda of agreement with the NIH. So much of our S & T has multiple uses that we invest heavily in partnerships to maximize return and benefit for the taxpayer. Currently, interdisciplinary research drives science, which means the formations of new partnerships with diverse people and institutions. One of NASA's programs, "Mission to the Human Body," partners with the National Cancer Institute (NCI). This program focuses on developing technologies that can detect, diagnose, and treat disease using nanosensors—detecting changes at the cellular level, even from within a cell. Conceptually, NASA wants to be able to detect cell changes that might lead to cancer before it becomes malignant, or similar types of strategies for different types of diseases.

NASA is interested in this technology because there will not be any emergency rooms accessible on Mars, or even on the International Space Station. We therefore must learn to prevent injury and diseases and revolutionize our health care delivery for the safety of the astronauts. The technology is twenty years away, but we already have teams of engineers, biologists, clinicians, and others starting to develop molecular sensors and information systems that will enable the medical care revolution. With both agencies excited about the prospects and willing to work together in this exciting area, we have managed to convince the appropriators to provide new funding to support peer-reviewed research proposals.

Although new funding for research programs has not been easy to attain in recent years, another area where NASA has been successful is a research program called "Living with a Star," investigating the sun-earth connection. This program emerged from a bottom-up proposal from scientists coming together from different agencies and within academia to look at priorities for space research. The program is a partnering of NSF, DOE, and the National Oceanic and Atmospheric Administration (NOAA) to develop the ability to study the sun and the effects of the sun on the earth.

Finally, I should mention NASA's recent emphasis on biology, including the promotion of exploration biology, astrobiology, health research, fundamental biology, the search for life, biology inspired technologies, and earth's carbon cycle.

The emphasis on new and exciting discoveries can actually be a big problem in the research world. Sometimes there is great value in ongoing research, or there are measurements that may need to be made continually over decades. It is a challenge to keep that required level of excitement and urgency when the research is not new and different. The problem is even worse for infrastructure. Universities are saying that laboratories are crumbling, new facilities need to be built, and equipment is old and inadequate. These concerns are critical—absolutely essential—but are not exciting.

THE UNIVERSITY–FEDERAL AGENCY RELATIONSHIP

I now want to examine the relationship between federal research agencies and their university partners. Universities are increasingly becoming interdisciplinary. Some parts of the university infrastructure have to adapt, like promotions and evaluation for tenure, as people move out of their departmental stovepipes and work in interdisciplinary teams. When federal research money is invested in interdisciplinary research, it is an amazing motivator. Universities are also partnering more with other universities and colleges, as well as with industry, state and local governments, nonprofit organizations, and multiple other federal research agencies. Universities receive more than 60 percent of NASA grants. These complex relationships are necessary in today's changing research environment, but they require new management strategies and much coordination and joint planning.

Federal agencies are driven by their mission, to advance S & T, ensure national security, and sustain the economy. One of the key roles of the university within these goals is to provide a skilled workforce. Although university researchers are more purely science driven, they also have the critical task of training and educating the S & T workforce of the next generation and ensuring an educated populace. This is an area where we all need to do some strategic thinking. The trends in the United States right now are frightening. There are hundreds of statistics, but figure 8.3 pictorially shows why we are concerned. The twenty-first century is supposed to be an amazing century of S & T—the knowledge-based society—but where will we find all the new engineers and scientists that will be needed?

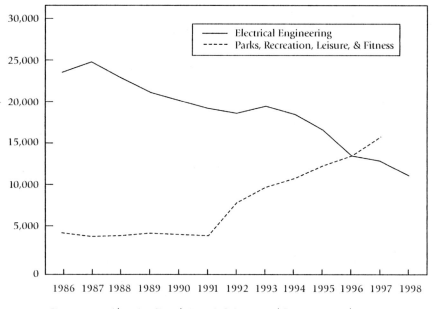

Figure 8.3. Alarming Trends in U.S. Science and Engineering Education

Universities do have influence with the federal government, especially with Congress. Their researchers do the bulk of federally funded research, as well as train the next generation of professionals, so they are the major drivers of the S & T budgets. For this reason, federal administrators are eager to work with their university partners to explain programs and identify opportunities for peer-reviewed grants. In this partnership, then, the role of the university is to know the program, the agency program manager, and the political climate. Their role is also to get involved in the program as a reviewer or an advisory group member.

University researchers need to know the roles of the agencies and their members. There are multiple opportunities, and each requires a different strategy. As an academic researcher, I worked in a laboratory carrying out neuroscience research. I thought of the NIH as the "shopping mall" of neuroscience research opportunities. I would look at which institute had the best "sale" on (e.g., the best funding rate, the highest average award size, and the longest duration). Back in the 1970s, the Eye Institute was providing funding at a very high success rate, but I studied gender differences in the brain. In my lab, we would joke about proposing some fundamental connection between vision and gender so that we could try for funding where the chances were better. But, although there are multiple opportunities and one tries to target research to the well-funded ones, ultimately a scientist's research program must be appropriate for the agency's mission.

Ensuring Quality

Federal agencies also must implement strategies to ensure the quality of the research they fund. Although the strategies are generally the same across all the research agencies, I will focus on the NSF.

The NSF has the responsibility to think about and advise the nation's overall capability in research, the future of research, and how we get where we want to be. There are four core values in NSF's strategy: peer review, stimulating high-risk/high-payoff research, encouraging interdisciplinary research, and education. Peer review, combined with merit competition, is the standard for the selection of federally funded research grants. This philosophy was articulated by Vannevar Bush in his 1945 paper, *Science, the Endless Frontier*, and has remained the hallmark of the U.S. federally funded research program.

Peer review assures excellence. But, alone, it can lead to programs that are risk adverse. Truly innovative ideas are often controversial and may be doubted by the established community. This is why NSF has an additional emphasis on stimulating high-risk—that is, innovative—research. NASA, too, has decided to sponsor some solicitations specifically seeking new and innovative ideas. Especially in technology programs, NASA has sponsored solicitations where evaluation is heavily weighted toward innovation and risk. For these opportunities, concepts already in latter stages of development or that are simply derivative in nature are not even considered.

The intent to stimulate interdisciplinary research was behind the creation of the NSF's S & T Centers (STC), a concept developed in 1987 to facilitate the exploration of new research areas and build bridges among disciplines, institutions, and other sectors. It is my firm belief that the breakthrough research of the coming decades will occur at the intersection of various disciplines.

The STCs also emphasize education and training. NSF insists on an education component in all of their research grants, and NASA is beginning to follow that lead. The development of our future professional workforce must be more integrated into our current research activities if we expect those activities to continue producing high-quality work in the future. In all the federal research agencies, we are keeping more statistics on the number of students and postdoctoral fellows who are funded, their stipends and working conditions, and their training opportunities and professional development. From there, we will assess whether we have the right opportunities and working conditions to attract more of the "brightest and best" into research. Herein lies our future.

CONCLUSION

The final picture for federal funding of R & D is that decisions are always going to be public and always political. There will always be concerns beyond those of expert members of the scientific community who advise us on the content of federal research programs. There are times when we scientists would like to live in our ivory towers, but our science agencies are part of an administration that has an established research agenda, and they are supported by elected officials who have local and regional concerns and need to balance the demands on the federal budget, as well as balance the budget itself.

However, we federal science managers work to make the scientist's case heard and to curb "pork" when we feel that it conflicts with the conduct of our agency mission. At the same time, we know that this tension will always exist. That is why it is so important for you to understand how the budgets are made and how agency decisions are made. We accomplish more when we work within the system rather than just trying to fight against what we will never change. We all agree that the national investment in S & T is critical, and this funding needs to be balanced and based on maximizing science return for society. As we have seen, funding priorities change over time, but all research disciplines contribute to our knowledge base.

Along with investing in S & T, we must invest more in the future generation of scientists and engineers, and also in the decision makers who will need to know something about S & T to make informed decisions, and this includes the general public. Today's important policy questions concerning stem cell research, genetically modified foods, and climate change all require informed choices. The number of similar technology-based challenges in the future will only increase. The trends in public science literacy and in students choosing science or engineering as a career are not promising. Thus, to accomplish our joint priorities, the partnership between the federal funding agencies and the university research community must remain strong.

Following are Dr. Olsen's and Dr. Carlson's responses to questions posed during a discussion of their paper at the University of Illinois Center for Advanced Study.

Q: You use the phrase "return on investment," which in application to a return in R & D seems to be a dangerous metric. Can you tell me how this has impacted NASA's budget requests and how much time NASA devotes to the metric?

A: *When I was at NASA, I probably devoted 10 to 20 percent of the year to those metrics. The GPRA is bringing this home. We cannot propose anything that does not have a demonstrable return. What is the metric? What is the outcome? How are we going to measure it? I have organized a recurring meeting with all these agencies, NASA, NSF, NIH, DOE, DOD, and others, coming together because all of us have a hard time addressing this. But the reality is that we are all receiving more and more pressure. We are trying to educate Congress on the importance of basic research—fundamental research. Although universities, and even industry, need to play a role* in basic research, they cannot proceed alone in their endeavors; the bottom line is that the federal government needs to support fundamental research and innovative research.

Q: Earlier you said that one role of the federal government is to help the economy, as well as helping society, in instrumental, tangible ways. Yet several times in your talk you referred to items that I would refer to as having more intrinsic value, that is, the search for life, where are we going, where have we come from, these sorts of things. How can you resolve those two when the stated value is instrumental and yet there's this sort of more intrinsic driver that you've cited several times?

A: *Basically in the government right now the drivers are security, economy, well-being, and education—those are the major players. The way that we sell things like the search for life is by hoping to find something that will help us cure a disease, or because studying these exciting questions helps draw students to the study of science. There is a place in the government for curiosity-driven research. The way that we sold the Origins program, which is astrobiology, and underlying that, the search for life, was by getting kids hooked on science and by getting the United States interested again in investing in NASA for the excitement of discovery; we successfully sold it on that education, enthusiasm line, which was the only way that we could.*

9

The Ethical Challenges of the Academic Pork Barrel

James D. Savage

Academic pork barreling is all the rage in higher education these days. Although the practice of academic pork barreling or academic earmarking has existed since the late 1970s, the current growth rate in earmarking is unsurpassed. In fact, the *Chronicle of Higher Education* reported that in FY2003, more than $2.012 billion in federal research funds were pork barreled for more than seven hundred universities and colleges (Brainard and Borrego 2003). This figure represents a 10 percent increase in pork barreled funding from the previous year, a far greater rate of growth than for traditionally funded academic research sponsored by any of the federal science agencies. While Congress has earmarked at least $13.5 billion for universities and colleges between FY1980 and FY2003, approximately half of that total was appropriated in just the last four of those years. The 2003 figure is more than six times the level of earmarking in 1996. Academic earmarking is now a widespread practice expanding at a very rapid rate. Yet what I would like to suggest is that beyond the matter of its cost and rate of growth, academic pork barreling presents higher education with a profound ethical challenge, forcing university leaders and faculty to choose between their personal and institutional self-interest and the general interest and well-being of the academic science community.

ACADEMIC PORK BARRELING: DEFINING TERMS

So what exactly is academic pork barreling, and why is it creating such a fuss? During the last twenty years, a growing number of universities and colleges have elected to bypass the regular peer- and merit-review processes

129

that evaluate applications for federal research funding in favor of direct political action. They do so by soliciting members of Congress, often by way of hired lobbyists, to "earmark" one of the government's thirteen appropriations bills, which fund the various federal research programs and science agencies, among other things. An earmark, then, is a legislative provision that designates special consideration, treatment, funding, or rules for federal agencies or beneficiaries. In a single sentence or less, the earmark orders an agency to award from its budget "$1,000,000 to Columbia University for its audubon [sic] research project," "$1,000,000 for the University of Akron for nanotechnology," "$1,000,000 for Montana State University for the tech link program," and "$1,500,000 for the Santa Ana College Space Education Center."

Virtually all academics know of Columbia University, and others might be familiar with the University of Akron and Montana State University. Only a very few, I suspect, are aware of Santa Ana College, a junior college in California, with no upper-division or graduate students or research faculty. Nevertheless, without regard for the recipient institution's academic standing or ability to support a meaningful research effort, each of these four earmarks was funded from the FY2001 Veterans Administration, Housing and Urban Development, Independent Agencies appropriations bill, with the money coming directly from NASA and the Environmental Protection Agency (EPA) budgets. In addition to these direct charges to NASA and EPA, the other major science agency funded from this bill, the NSF, also suffers from this type of earmarking, even when the earmark is not directly funded from the agency's accounts. Under the rules of the federal budgetary process, each appropriations subcommittee receives a designated allocation of money to spend. So the cost of these earmarks potentially reduces the amount of funding available for all other programs supported by the subcommittee, including the NSF, in a zero-sum fashion.

Reasons for the Rise in Academic Earmarking

The long-term trend in earmarking is clearly upward, subject to intermittent periods of apparent restraint. Academic earmarking progressed steadily from 1980 through 1993 and dipped in the mid-1990s, only to accelerate rapidly since 1999. There are several reasons for this recent and dramatic rise in pork barreling. The federal budget during the late 1990s was running surpluses rather than deficits. So in FY1999 and FY2000 the Republican-controlled Congress, with the urging of the Democratic minority, exceeded the spending caps enacted in the 1997 Balanced Budget Act by tens of billions of dollars. Historically, there is a direct correlation between the size of discretionary spending available for the appropriations committees, known as the 302a allocations, and the amount of earmarking. During the early and mid-1990s, when the desire to control deficit spending reached its peak, the tight spending caps imposed by the 1990 Budget Enforcement Act constrained these appropriations; hence the reduction in academic earmarking.

Another reason for this growth was the Republicans' embrace of earmarking in 1995. In 1994, the Republicans won control of the Congress, and, particularly in the House of Representatives where Republicans proclaimed the virtues of the "Contract with America," they swore they would control pork barreling and excessive spending. In one famous incident, Robert Livingston (R-LA), the newly appointed chair of the House Appropriations Committee, called a meeting of the full committee, brandished two large alligator skinning knives, and announced that he was going to "hunt pork like Louisianans hunted gators."

Indeed, earmarking did fall significantly from 1993 to 1994. When it came to pork-barrel spending, however, the contract only lasted about nine months. In July 1995, Speaker Newt Gingrich (R-GA) sent a memo to the appropriations subcommittee chairs, asking them, "Are there any Republican members who could be severely hurt by the bill or who need a specific district item in the bill?" (Savage 1999, 137). With that encouragement, earmarking would again soon be on the rise, as Richard Munson, executive director of the Northeast-Midwest Institute correctly predicted in 1996. "I bet you're going to find a big bump up in FY1997, so that you're recording your low point now," Munson claimed, as the Republicans elected in 1994 would realize "the benefits of incumbency" (Cordes and Gorman 1996, A33). Academic pork barreling is not limited to party divisions but is in fact a bipartisan activity.

The late 1990s also witnessed a diminished number of congressional "saints" who championed the fight against earmarking. In political science jargon, saints are members of Congress who, as a matter of principle, fight against pork-barrel projects of all types despite the threat of reprisals by peers and the denial of reelection-winning projects for their own districts (Savage 1991). Such saintly members for a variety of motivations oppose academic earmarks, and during the 1990s they included representatives William Natcher (D-KY), George Brown (D-CA), Edward Boland (D-MA), and Robert Walker (R-PA) and senators William Proxmire (D-WI), Sam Nunn (D-GA), John Danforth (R-MO), and Jake Garn (R-UT). Natcher and Brown passed away, and the others retired, leaving Senator John McCain (R-AZ) as the most prominent member in active opposition to academic earmarking.

Earmarking also expanded during this period due to what might be called a "contagion" or learning effect throughout higher education. Over the years, university presidents and their faculty learn from their peers, and from the high-powered sales pitches of lobbyists who specialize in earmarking, just how to go about securing such projects. These presidents and faculty not only learn the mechanics of earmarking, but also that there are no sanctions imposed on institutions for engaging in the practice. Despite the official pronouncements of numerous academic associations and scientific societies, no academic institution, president, or faculty member has ever been sanctioned in any form for participating in earmarking. This absence of sanctions, combined with the endless examples of virtually every major university in the nation benefiting from earmarking, simply encourages more and more schools to seek earmarks.

Significantly, a number of these new earmarking institutions make no pretense about their capabilities as research universities. These include junior colleges as well as four-year institutions with limited graduate programs. If the leadership of research-oriented universities believed earmarking would only suit their needs, they were truly mistaken.

One interesting example of the devolution of earmarking to non-research-oriented institutions is the growth of earmarking by California State College in San Bernardino, a school that is constrained from issuing doctoral degrees by California's Master Plan for Higher Education. Yet, despite this, the school has obtained almost $20 million in earmarks and plans on securing an additional $25 million in the next few years with the aid of a Washington lobbyist hired in 1999. Included in the $20 million worth of earmarks is a $5 million distance-learning project funded from the Navy's budget and $3.7 million for a technology transfer center. These projects along with others were earmarked with the assistance of that area's member of Congress, Representative Jerry Lewis (R-CA), a powerful appropriations subcommittee chair. Since he became a chair, Cal State has assiduously courted Lewis, asking him to even be a commencement speaker.

In one of the great ironies in the brief history of academic earmarking, Cal State has proposed that its plans for an earmarked math and science building valued at $19.5 million be named after Representative George Brown, the deceased chair of the House Science Committee who so bitterly fought against academic earmarking. Brown's widow, Marta Marcias Brown, quickly requested that the building not be named for her husband, a request Cal State agreed to honor (Tresaugue 2001a, 2001b).

Arguments Supporting Earmarking

Although earmarks are growing in number and in their total dollar value, advocates of earmarking offer four familiar, and sometimes compelling, arguments to defend their projects. First, earmarking serves as a remedy for the biases of the dominant peer-review system. John Silber, the former president of Boston University and the most articulate advocate of earmarking, makes the claim that peer review is an "old boys" club, where the same scientists fund their friends and colleagues from the same universities doing the same research. "The system of distributing funds by peer review rather than congressional discretion had the effect of creating a flawed system, where only the members of the in group were in a position to receive funds" (Silber 2001).

Peer review is inbred, self-sustaining, and exclusionary; gaining entrance to the club is an exceptionally difficult task. Consequently, to permit more equitable and fair access, those faculty and institutions traditionally excluded should benefit from the extra financial support stemming from earmarking. Earmarks fund the new buildings and research projects that will attract high-quality faculty, who will then allow the school to compete for the peer-reviewed funds, and in Silber's words, "create new centers of scientific excellence."

Second, earmarking offsets the absence of meaningful federal research facility programs. The federal government has largely failed to support the construction and maintenance of academic research facilities, apart from the inadequately budgeted programs in the NSF and NIH. The brief lifespan of the NSF program, for example, was terminated during the Clinton administration, with a total of only approximately $230 million in appropriations over a period of seven years. In this way, the federal government has reneged on its "social contract" with higher education, where the federal government would fully fund the costs incurred in the conduct of research by the higher-education community. Meanwhile, the backlog of deferred construction of various research facilities is measured in the billions of dollars. This backlog is particularly problematic for those institutions aspiring to break into the "old boy" peer-review club. Earmarking is a way to gain the funding for facilities needed by all members of the academic science community, especially so that "emerging" or "developing" institutions can compete with their elite counterparts.

Third, earmarking also acts as a remedy for the inadequacy and bias of indirect costs. As is well known, the direct-cost component of a research grant is spent by the principal investigator to conduct research, and the indirect-cost component is the institution's share for supporting those research activities. Thus, indirect costs act as a reimbursement for the institution's providing libraries, staff support, building maintenance, and so forth. One reason for the virtual elimination of funding for research facilities is that during the late 1960s, research universities pressed the federal government to replace direct facilities funding with higher indirect-cost rates, which prior to 1968 were capped at about 20 percent of the value of the research grant. Indirect-cost rates spiraled up to rates that sometimes reached 60, 70, or nearly 100 percent of the value of the research grant. One of the elements of the indirect-cost rate attempted to measure facilities' expenses associated with supporting the research grant in question, and therefore it could be argued that the federal government was indeed fully funding research. Yet these high rates were usually only charged by elite private research universities, and in response, both Congress and many of the executive branch agencies attempted to recap or even negotiate these rates downward. More to the point for earmarking universities, indirect costs only help an institution if the institution received a research grant. Indirect costs essentially reward successful universities; relatively speaking, they provide little or no assistance to emerging or developing universities that are attempting to construct new research facilities. So, again, earmarking is said to offset the biases of the established peer-reviewed contract and grant system.

Fourth, a far less intellectually compelling but certainly powerful argument for earmarking is that of peer pressure, or in other words, because "everyone else is doing it." There can be little doubt that university leaders and faculty observe and learn from one another, so that if some schools are earmarking with impunity, then perhaps they should also. Moreover, the widespread violation of the various resolutions and moratoria that have been passed by various organizations such as the American Association of Universities (AAU), by the very member institutions that voted for them, confirms the idea that "everyone else is doing it." Recently, a lobbyist who specializes in academic earmarking told me that a prospective client, a newly installed university president, was hesitant about approving an earmark for the campus because of resolutions passed by professional organizations like the AAU discouraging earmarking. In response, the lobbyist handed that president a list of AAU universities that violated the AAU moratorium, a list that included virtually every member institution, and added the remark, "Everyone else is doing it." The earmarked project was then quickly approved.

There is an additional, fifth argument, pointing out that it is well within the rights and prerogatives of the Congress to earmark legislation. The Constitution empowered the Congress with the authority and duty to create appropriations, and nothing in the Constitution says members shall not aid their constituents with earmarks. This is, of course, absolutely correct, and many members argue that earmarking is good public policy and offsets the biases of peer review. Moreover, Congress never passed any moratoria or resolutions against academic earmarking; university presidents and academic and scientific societies did, and they are responsible for obeying and enforcing these codes, not Congress. Furthermore, although there are indeed entrepreneurial members who encourage academic earmarking, these projects overwhelmingly stem from universities and their lobbyists requesting earmarks from their senators or representatives. Though members of Congress have the right to earmark, they cannot be blamed for the rise in earmarking. The controversy over earmarking originated within academic associations and societies, and they are responsible for resolving that controversy.

Arguments Opposing Earmarking

Meanwhile, there are powerful arguments that can be raised in opposition to earmarking. First, earmarking undermines the peer-review system. There are a number of allocation systems by which governments distribute research funds to universities, and peer review is just one of these. In some Scandinavian countries—Finland, for example—funds are distributed on a student enrollment basis, so the larger the enrollment, the greater the funding, regardless of scientific merit. In England, funds are allocated through institutional block grants. In the United States prior to World War II, perhaps the dominant system consisted of formula funding to land-grant colleges and agriculture schools. As students of science policy history know, the dominance of this process shifted during World War II because of the federal government's reliance on project grants to meet this national (and global) emergency.

Toward the end of the war, Vannevar Bush pushed for the continuance of project grants over formula funding in his famous report to President Franklin Delano Roosevelt, *Science, the Endless Frontier* (Bush 1945). Bush's reasoning was twofold. He first sought to ensure that federal funding supported the highest-quality research and that this would be realized through a merit-based selection of individual projects, not by automatic formula grants that rewarded research regardless of quality. Bush, though he served in the wartime government, was an academic on loan from MIT, and by way of peer review he attempted to insulate academic science from the patronage politics of either parochial members of Congress or fief-building administrators. Peer-reviewed project grants enabled academic scientists themselves to determine where the money went. Although Bush's specific recommendations were never formally enacted into law, over time many of the most prominent science funding agencies, such as the NIH and the NSF, turned to project grants judged by peer-review panels.

Peer review is, of course, rooted in the scientific method and is the overarching decision-making process that influences virtually every critical decision in academic science and the academy. Peer review is the system that is used to determine tenure, promotion, publication, and a host of other evaluative procedures influencing academic life. When an institution or its faculty, for example, attempts to bypass peer review, as in the case of cold fusion, where University of Utah scientists declared cold fusion to be a reality prior to submitting their findings to peer review, it is often the case that they are subject to ridicule. Peer review is a fundamental characteristic of the academy, and its extension to the allocation of federal science funds, as championed by Vannevar Bush, extended academic norms to the otherwise political and bureaucratic patronage systems associated with federal spending. Declaring peer review to be an inherently and irreparably biased and flawed system undermines that precious barrier, separating academic science from overt political intervention and direction in the conduct of research.

Second, if peer review is presumably biased, concentrating resources in the hands of a few privileged universities, the same problems exist involving earmarking. Proponents of earmarking assert that one of the most telling bits of evidence proving the bias of peer review is that research funding is highly concentrated among a handful of institutions. Indeed, under the peer-review system, the top ten university recipients of federal research dollars receive 22 percent of total obligations, based on the annual compilation of such data by the NSF. The top twenty schools receive 37 percent, and the top thirty schools 48 percent. This leaves about half of the balance for the rest of the schools and universities. Consequently, so it is argued, earmarking serves to redress this imbalance and inequitable distribution of federal dollars.

Yet earmarking is also highly concentrated, reflecting the simple fact that neither scientific merit nor political influence nor power is equally distributed. Of the $8 billion earmarked between 1980 and 2000, about 60 percent went to schools represented by chairs of the powerful Senate and House appropriations subcommittees. These chairs are key members who set funding priorities within the thirteen appropriations subcommittees, and it is these members who receive their colleagues' requests for budgetary favors, including academic earmarks. While granting some of these requests, these chairs take the lion's share of these projects for their own constituents. Power is distributed unequally within the Congress, and thus so are its rewards. So, such familiar members of Congress as senators Daniel Inouye (D-HI), Ted Stevens (R-AL), Robert Byrd (D-WVA), Mark Hatfield (R-OR), and J. Bennett Johnston (D-LA) have propelled the universities of Hawaii, Alaska, and West Virginia; the Oregon Health Science University; and Louisiana State University into the top ten recipients of earmarked funds. Meanwhile, representatives John P. Murtha (D-PA) and Jerry Lewis (R-CA) accomplished the same for the University of Pittsburgh, Pennsylvania State University, and Loma Linda University.

This hierarchy is reflected in the funding of academic earmarks, where a handful of schools reap the benefits of these politically determined projects. Where the top ten universities received 22 percent of total federal research obligations, the top ten earmarking schools obtained 21 percent of earmarked funds as mentioned. Where the top twenty universities received 37 percent of total obligations, the top twenty earmarking universities secured 32 percent of total earmarks. Finally, where the top thirty schools received 48 percent of total obligations, the top thirty earmarking institutions received 41 percent of total earmarks. Thus, rather than providing a remedy for the concentration of research funding, academic earmarking produces its own biases and its own political "old boys club" concentrating research funding.

Third, earmarked funds are wasteful and ineffectively spent. A common argument against earmarking is that such funds are inherently inefficient; otherwise these projects, especially those that are explicitly for research rather than simply brick-and-mortar facilities, could be won through competitive review. Yet, as University of California president David Gardner pointed out,

"What facility is not meritorious?" (Savage 1999, 58). In other words, an earmarked project should provide an institution with some degree of value, even if a research facility is turned into a simple warehouse. Nevertheless, the issue here is not the worth of an earmark in some absolute sense, such that it provides any value to an institution, but its relative or competitive sense. Was this earmark the best way to spend taxpayer dollars, and, more specifically, were these funds effectively employed by the institution to make it a more competitive university?

This question is important because the claim made, for example by President Silber, is that earmarked facilities and research projects will attract higher-quality faculty to an institution, which will then be able to compete for peer-reviewed funding. "The recipients of earmarked funds," Silber states, "have leveraged those grants many times over and demonstrated their value through the scientific advances they have made" (Silber 2001). So, do earmarked funds act in effect as "seed corn," enabling institutions to improve their research capability and competitiveness?

Table 9.1 indicates the change in federal research rank among the top thirty-five institutions receiving academic earmarks during FY1980 through FY2000, where each of these schools obtained at least $47 million in appropriations. The table identifies the institution, the total dollar amount earmarked, the first year in which earmarks for that school totaled at least $1 million, the federal research rank that year, the rank in 1998, and the rank change. The research rank is based on the annual publication of total federal research obligations to colleges and universities published by the NSF, which ranks the top one hundred recipients.

What these data suggest is that the record of improved research capability through earmarking in a relative sense is, at best, very mixed. As of 1998, eight of these institutions increased their ranking, and eleven experienced a loss of ranking compared to their rank during the first year they received significant earmarked dollars. The remaining institutions could not be compared on a relative basis since schools receiving less than those in the top one hundred are not ranked by the NSF. Of the eight that increased their ranking, only six schools—Georgetown, Oregon Health Sciences, Rutgers, and Boston and Indiana Universities—boosted their ranking by five or more places, suggesting meaningful improvement. Meanwhile, seven institutions actually dropped five or more ranks, including the top beneficiary of academic earmarking, the University of Hawaii, which fell nine ranks despite being the recipient of nearly $300 million (Malakoff 2001, 835; Savage 2001). Thus, after challenging the peer-review system by earmarking $8 billion of taxpayer dollars, the net result is that only six institutions appear to have significantly strengthened their competitive research capabilities.

What makes this performance even more discouraging is that the NSF data do not discriminate between earmarked and peer-reviewed funds. So, for

Table 9.1. Change in Federal Rank among the Top 35 Institutions Receiving Apparent Academic Earmarks, FY 1980–2000 (Source: Federal Support to Universities, Colleges, and Selected Nonprofit Institutions. (Washington, D.C.: NSF). Various fiscal years (1980–1998).

Cumulative FY 1980–2000 earmark rank	$ Amount	Year of first $1 million earmark	Federal research rank that year	FY 1998 federal research rank	Rank change
1. U. Hawaii	$290,409,500	1983	70	79	-9
2. U. Alaska	198,764,572	1990	—	—	—
3. Georgetown U.	179,296,000	1985	93	67	+26
4. U. West Virginia	172,186,300	1991	—	—	—
5. U. Pittsburgh	162,164,198	1991	15	19	-4
6. Iowa State U.	161,962,000	1987	—	—	—
7. Louisiana St. U.	155,083,800	1983	60	53	+7
8. Loma Linda U.	150,600,000	1988	—	—	—
9. Oregon Health Sci. U.	125,474,696	1983	—	65	+35
10. Pennsylvania St. U.	97,835,000	1983	19	21	-2
Subtotal: $1,693,776,066 21% of cumulative total of $7.935 billion					
11. Oregon St. U.	97,341,333	1983	50	71	-21
12. U. Florida	95,997,700	1988	42	44	-2
13. Mississippi St. U.	95,083,100	1985	—	—	—
14. U. Mississippi	92,085,417	1988	—	—	—
15. Michigan St. U.	87,800,534	1988	49	57	-8
16. U. Rochester	81,139,396	1983	24	28	-4
17. U. Maryland	79,615,700	1990	41	46	-5

	Amount	Year			
18. U. Nebraska	75,307,600	1990	—	—	—
19. Rutgers U.	72,668,400	1980	76	60	+16
20. U. North Dakota	71,703,300	1986	—	—	—

Subtotal: $848,742,480 / $2,542,518,546   32% of cumulative total

	Amount	Year			
21. Washington St. U.	71,584,300	1986	—	—	—
22. North Dakota St. U.	68,354,633	1980	—	—	—
23. U. Scranton	62,300,000	1988	—	—	—
24. Tufts U.	61,924,300	1980	100	96	+4
25. Baylor Medical Coll.	61,602,000	1980	37	34	+3
26. Boston U.	60,500,000	1984	42	30	+12
27. Indiana U.	57,995,100	1986	49	40	+9
28. Sagniaw Valley St.U.	55,470,000	1990	—	—	—
29. U. Illinois	53,673,900	1986	17	27	-10
30. U. South Carolina	53,490,000	1986	—	—	—

Subtotal: $606,894,233 / $3,149,412,779   40% of cumulative total

	Amount	Year			
31. U. Massachusetts	52,565,000	1986	—	—	—
32. U. Wisconsin	52,478,600	1989	9	15	-6
33. U. Miami	49,435,600	1990	42	49	-7
34. U. Oregon	49,239,300	1987	—	—	—
35. Florida St. U.	47,544,100	1985	88	88	0

Subtotal: $3,400,675,379   43% of cumulative total

ranking purposes, these institutions benefit from both types of funding. Yet institutions such as the University of Hawaii, the University of Pittsburgh, Pennsylvania State University, and Oregon State University, which have received a combined total of nearly $650 million for projects of their own choosing, have failed to maintain even a steady-state ranking. Moreover, the overall performance of earmarking universities has declined with time. When a similar analysis was conducted based on 1994 data, thirteen schools increased their ranking and ten experienced a decline, versus only eight schools increasing in rank and eleven declining based on the 1998 data. Unfortunately, because the NSF data combine both peer-reviewed and earmarked data, and because the growth in earmarking is so dramatic, future analyses of this type may be highly suspect, as true institutional performance will likely be masked by these earmarked dollars.

Fourth, earmarking creates an opportunity cost, thereby crowding out peer-reviewed projects. Although the exact amount would most likely be impossible to determine, it is reasonable to assert that an opportunity cost exists when earmarked funding comes at the expense of peer-reviewed funding. For example, in the original case that touched off the earmarking controversy in 1983 that resulted in the passing of the various earmarking moratoria and resolutions, Columbia University and Catholic University received $10 million in earmarks at the expense of Yale University and the University of Washington, which had won these funds in peer-reviewed competition. In addition, several of the science agencies have complained that earmarking diverts support from peer-reviewed programs, making the setting of agency science priorities extremely difficult. Even if no more than, say, a fourth of earmarks diverted funding in this manner, that means $2.5 billion of the nearly $10 billion in earmarked dollars could have been competitively granted to university researchers.

Fifth, earmarking promotes hypocrisy in the academy and in the academic sciences. During the mid-1980s, the following organizations formally opposed earmarking in some form: the Association of American Universities, the National Association of State Universities and Land Grant Colleges, the American Association for the Advancement of Science, the American Physical Society, the Federation of American Societies for Experimental Biology, the Council of Scientific Society Presidents, the Council of Graduate Schools, the American Association of State Colleges and Universities, the National Academy of Sciences, and the National Science Board.

The AAU, the association representing the interests of the major research universities, has in fact twice adopted resolutions against earmarking. The first resolution came in the form of a moratorium on earmarking in 1987, which was followed in 1989 by the Resolution on Facilities Funding, which requested that its membership abstain from earmarking. In the ensuing years, virtually every member of AAU violated these prohibitions against earmarking. They did so either overtly and with little hesitation, or in some more notable cases member institutions creatively interpreted the resolutions in a manner they claimed justified their earmark.

Universities have formed consortia where the earmark would be distributed among the members on the basis of "peer review," such that the peer-review panels would consist only of the faculty in the consortia, who then would determine which of the member institutions received what amount of funding. Universities have had legislative language added to appropriations that called for peer review in the allocation of research funding for certain types of projects, but this same language limited the institutions that might compete for the funds to only those that were predetermined to benefit from these programs. Universities received earmarks, had the project evaluated after the fact by the scientists engaged in the project, and then claimed that this was peer review.

In one particularly interesting example of redefining an earmark, Harvard University, the institution often offered as the standard of excellence in American higher education, whose school motto is *Veritas*, or "Truth," determined that it would not openly seek an earmark. The university, however, then did leave itself a loophole, saying that if another institution received earmarked funding and wanted to share some of those dollars with Harvard, that was acceptable. As President Derek Bok noted in a letter to a Harvard faculty member,

> I believe that Harvard should not try to avoid the regular peer review procedure to evaluating the merits of scientific proposals. Hence, I would be very concerned if representatives of the University solicited members of Congress to earmark funds for us as line-item appropriations. On the other hand, if money is given to another university which wants us to collaborate in a worthy scientific or research enterprise, I would have no objection. (Savage 1999, 67–68)

In other words, if another institution did the political dirty work and secured the project, Harvard would gladly share in the benefits of the earmark.

The university presidents who voted in favor of these various resolutions have completely failed to enforce their own pronouncements against earmarking. Not a single institution, president, or faculty member has been penalized for earmarking. No scientific society or national association, for example, has terminated the membership of a transgressor. Meanwhile, returning back to the AAU, membership has actually been extended to such institutions engaged in significant earmarking as Rutgers University, the University of Arizona, the University of California at Santa Barbara, and the University of Florida. Rather than discouraging earmarking, AAU has encouraged it by rewarding earmarking institutions with its greatest honor, membership in the organization.

THE ETHICAL CHALLENGES OF EARMARKING:
WHAT CAN BE DONE?

Academic earmarking is partly a story of science funding and budgeting, of research facilities needs and indirect costs, of lobbyists and powerful members of Congress. More important, however, is the story of how the leaders of higher education have responded to the ethical challenge of honoring their own resolutions. For when the AAU presidents and the various associations and academic societies declared earmarking to be a problem, they declared that it was more than a budgetary issue; it was an act that violated long-standing norms and the core values of academic science. So the ethical challenge thus presents itself whenever a president decides whether to hire a lobbyist and seek an earmark.

For some university presidents, such as President Silber, earmarking poses no such ethical dilemma. It is simply regarded as a welcome remedy to the biases and limitations of peer review. Ironically, though, even Silber, advocate of earmarking that he is, measures his own institution's accomplishments by the standard of peer-reviewed publications and federal research grants competitively won. Other presidents have found the decision to earmark a more troubling one. When former president Sheldon Hackney of the University of Pennsylvania finally elected to hire a lobbyist and pursue an earmark, he recognized the conflict between that decision and his votes at the AAU. "During that time," Hackney noted, "I certainly recognized my responsibility as a citizen and a university president to think and act in the public interest, but I also have a responsibility to my institution to do what's best for it" (Savage 1999, 59–60).

There is a cost to such a choice. The decision of some presidents to violate the very resolutions and moratoria they or their institutions voted for leaves observers skeptical as to the sincerity and integrity of these choices. Meanwhile, such decisions encourage other institutions to seek earmarks under the justification that "everyone else does it." The willingness of these presidents to violate academic norms is further seen by some as having the potential to infect the entire scientific process. As one vice president for research at a major university privately commented, "When we train our young PhDs, does this mean that we have to train them in the political process of how to get earmarks so that they can fund their research?"

The cynicism that earmarking creates may be summed up in an article appearing in *Prism*, the magazine of the American Society for Engineering Education. "On the one hand, university presidents and academic organizations condemn the practice on a theoretical level, but on a practical level, they hire lobbyists, chase down the money, fawn over politicians, and cash the checks" (McGraw 2001, 25). Thus, university presidents, who serve as the ethical leaders of their faculty and student bodies, may appear no better than shady political operatives, even to sympathetic observers.

One way of resolving the earmarking problem in a formal sense, of course, is for various professional academic associations, particularly the AAU, to revisit the question of whether academic earmarking is an acceptable practice. A new round of voting on these resolutions might produce a different set of standards that affirm earmarking as a legitimate, parallel system of funding academic science. That way, at least, those institutions that once voted for the resolutions against earmarking would no longer be acting in openly hypocritical manner. The danger of such a new position, however, is then there truly would be no constraints on earmarking, which would leave all of academic science with little choice but to engage fully in the politics of research funding. Under these conditions, as that vice president for research feared, graduate students would indeed have to be trained in the art of influence peddling.

If, on the other hand, these associations elected to reaffirm their stand against earmarking, a reinvigorated opposition to earmarking could include the following activities:

- Bring together congressional and academic leaders who oppose earmarking, so they might reinforce each other's positions and coordinate their activities. Regular meetings are truly rare occurrences and usually are small-scale affairs. Perhaps more members of Congress would become "saints" who opposed earmarking if they were aware of sympathetic colleagues. I cannot recall there ever being a meeting that brought together House and Senate saints and university presidents to collaborate on a strategy to oppose earmarking.
- Educate members, especially those on appropriations committees, about the value of peer review and the harm of earmarking. Recruit new congressional saints.
- Use empirical data that show the limited effect of earmarking in improving institutional capabilities, and the distributional biases of earmarking. Mobilize constituent universities and faculty to encourage their members to oppose earmarking.
- Reward members of Congress and university presidents who oppose earmarking. Members who engage in earmarking often receive honorary degrees and get buildings named in their honor. These types of honors, banquets, and prizes should also be offered in recognition of those who actively oppose earmarking. Neither George Brown nor William Natcher was honored by any of the academic societies that formally opposed earmarking, but such prizes and honors could be made in their names. University presidents who make tough policy choices against earmarking also deserve public support and recognition from their peers.

- Presidential opposition to earmarking should be recognized and supported. The last three presidents of the United States have been particularly active in denouncing academic earmarking. President George H. W. Bush attempted to eliminate earmarks by way of a budgetary rescission. President Bill Clinton employed the line-item veto to delete earmarks. President George W. Bush's first budget denounced earmarks and proposed that several be denied funding. The academic science community was deathly silent in its support of the Clinton line-item veto and failed to endorse George W. Bush's position against earmarks.
- Enforce sanctions against presidents, faculty, and institutions that violate referenda and moratoria. Until academic science is willing to defend its own rules and values, there is little reason for Congress to limit its earmarking. The National Academy of Sciences, the American Academy for the Advancement of Science, the American Physical Society, and the AAU must be willing to sanction members, for example, by denying them membership, by changing membership criteria so that earmarking inhibits membership possibilities, and by publicly publishing lists of transgressing members, as the American Association of University Professors has been willing to publish lists of universities that violate its standards.
- Universities should enforce their faculty codes of conduct. It is not uncommon for university presidents to place the blame for earmarking on their faculty. Recently, the former president of an elite Big Ten university asserted that earmarking at his institution was due to faculty seeking earmarks by making direct contact with members of Congress. When asked if his faculty code of conduct prohibited such independent action, the president indicated it did, but the code was not enforced. Rogue faculty, for example, who initiate earmarks could be required to return their earmarked funds to the government.
- Higher-education associations and academic societies should do more to mobilize faculty. Faculty members have never been asked to write letters to their members of Congress to register their opposition to earmarking.

Obviously, some of these recommendations also carry the onerous task of sanctioning colleagues, which many will no doubt consider an unpleasant, if not impossible, task. Yet, at the present time, those who engage in academic earmarking do so with impunity. Resolutions and moratoria are generally ignored, and short of having the name of one's university published in the annual listing of pork-barrel projects in the *Chronicle of Higher Education*, there are no sanctions imposed on earmarking schools. What these suggested penalties and rewards might provide is some sort of breakwater against what is now a rising tide of academic earmarks.

Following are Dr. Savage's responses to questions posed during a discussion of his paper at the University of Illinois Center for Advanced Study.

Q: When one looks at how this trend has developed over time, can you make some comparison between the amount of money that is distributed with the peer-review process and how much money comes to universities as a result of earmarking?

A: *The total amount of federal obligations for academic research the last few years has been about $18 to 19 billion, and the pot of earmarked appropriations was nearly $1.7 billion in FY2001. So about 10 percent of these dollars are earmarked. This issue, however, is more than just about dollars and cents. Earmarking goes to the heart of what research universities are about, in terms of whether they are conducting research in a manner consistent with the scientific method, and in terms of whether university leaders and faculty behave in an ethical manner. The total of the earmarked dollar amount is not large, but in some agencies it is very high—for example, in the Department of Agriculture. There appears to be some crowding out of quality research by the level of earmarking, as reported by the federal agencies themselves. One more thing about the amount of earmarking: The total dollar value of earmarking that can be identified is about eight billion dollars in the last twenty years. When stretched out over time, that is not a lot of money, but consider, as I mentioned before, that about half of that has taken place in the last six years. Earmarking is really a growing phenomenon, since many more schools are participating in this practice. Earmarking raises the question of quality control in the allocation of research dollars. Getting an earmark in an appropriations bill is completely a matter of political influence, having little or nothing at all to do with academic quality. Occasionally there's sort of a pro forma attempt to deal with this issue of quality control. When Congressman Durbin (D-IL) was chairing, I believe, the Agricultural Appropriations Subcommittee in the House, he had a one-page sheet that was used to try and identify some of these concerns about quality, but by and large it was just paperwork, and brief paperwork at that. Getting an earmark is completely a matter of who you know.*

Q: Some people might say, "Well, at least the money is going to go somewhere. They have to give that money away; they can't keep it." Does this type of attitude have negative consequences for the government bodies that fund research?

A: *The federal agencies have been unanimous in opposing earmarking. However, it can be very dangerous for them to oppose earmarking too vocally, because they can run afoul of very powerful members of the Congress and could jeopardize their funding. One of the reasons these agency directors oppose earmarking is that it undermines their capacity to set priorities. What you are talking about here is the contrast between agencies creating programs that advance national priorities and promote good science, and the parochial desires of members to help their constituents. What happens is that the agencies submit budgets through the budgetary process which reflect their priorities, but they then find out later that a fair chunk of their money has been allocated for questionable projects. So it undermines their ability to set priorities, and then what happens is they end up supporting questionable research projects, which the agencies find distasteful.*

Q: From what you have described, there does not seem to be much accountability concerning earmarked funds. How is it that earmarked funds, as in the case with the University of Hawaii, could be given to a university with no conditions or expectation of return on the investment(s)? Because it seems like you are saying that earmarked funds do not have any specific purpose attached to them. And is this really that common?

A: *Schools that receive earmarks basically have that money granted to them with no strings attached. The university, not the funding agency, selects the project that will be funded. So the process is not one where the agency says, "Here are the national needs; here are established programs that we think need to be supported; here is the research that needs to be conducted in these areas in order to meet national needs." The projects, whether buildings or individual research projects, are selected by the universities themselves, so in essence, it is money to fund their wish list. It is the equivalent of giving universities a blank check, with no strings attached.*

Q: Going back to an earlier point, given the fact that there are many research institutions that are competing for funds and only so much money, this has led to more partnerships between private business and universities. However, I am also sure that this has led universities to seek out other alternative forms of funding, like foundation grants and so forth. While I am sure that we will continue to see federal money in research, do you think we will continue to see the federal government becoming less and less important in funding research in the future?

A: *If you look at the totals, the federal government is still the overwhelming provider of university research funds, and it will always be that way. You may see more of the private-public activities going on—by that I am referring to private-university activities—but these opportunities will not be available to all schools. There are some schools that have the capacity to generate these kinds of relationships, but there are other schools that are not going to be able to do that for all kinds of reasons. Historically, there has been sort of a split between what the private sector is willing to support for economic reasons versus the public sector. The reason why conservatives, for example, are willing to support university funding is because they understand that the private sector does not want to take all the economic risks that are associated with basic research. Basic research does not promise automatic product spin-offs and commercialization, as might applied and development research conducted in the private sector. So it is clear, as you are suggesting, that these public-private relationships are growing and becoming increasingly important, but you are not going to have a major research university of the stature of the University of Illinois, for example, without a lot of federal money involved.*

Q: Isn't it a fact that a lot of the scientists and the research groups have their funding denied simply because their work may no longer apply to armament or war productive industries?

A: *Actually, I think if you look at the sources of federal money in terms of the agencies, the Department of Defense (DOD) is still the third or fourth largest source of academic research support. Currently, the largest is clearly Health and Human Services (HHS) and the NIH, second would be the NSF, but third is the DOD. So universities get a lot of money from the DOD.*

Q: Is anyone addressing earmarking from this perspective: that is, if we consider ourselves a research university, how is it that we measure our worth based on how much money we can bring in and spend? The issue then becomes that we need to establish other ways to determine our worth and to talk about what it is we do, and maybe doing research at that level should not be the way that we do that.

A: *I think your point is well made, but it is also a fact that part of the problem is a proliferation of ambition, where so many academic institutions now want to be research oriented for reasons of status and money. Many of these schools are, at best, marginal and should not have any business viewing themselves as research institutions. When these schools view themselves this way, they create a new set of incentives that influences the rewards system, including tenure and promotion, which creates more pressure for an institution to seek earmarks.*

REFERENCES

Brainard, Jeffrey, and Anne Marie Borrego. 2003. Academic pork barrel tops $20-billion for the first time. *Chronicle of Higher Education*, September 26, A18.

Bush, Vannevar. 1945. *Science, the endless frontier: A report to the president on a program for postwar scientific research.* Washington, DC: U.S. Government Printing Office.

Cordes, Colleen, and Siobhan Gorman. 1996. Congress slashes earmarks for academe by 50%, the biggest decrease ever. *Chronicle of Higher Education*, September 13, A33.

Malakoff, David. 2001. Perfecting the art of the science deal. *Science* 292 (5518): 831–35.

McGraw, Dan. 2001. Playing the game. *Prism* 10 (7): 22–26.

Savage, James D. 1991. Saints and Cardinals in appropriations committees and the fight against distributive politics. *Legislative Studies Quarterly* 16 (3): 329–47.

———. 1999. *Funding science in America: Congress, universities, and the politics of the academic pork barrel.* New York: Cambridge University Press.

————. 2001. Nosing for cash in Congress's trough. *Times Higher Education Supplement* (London, England), March 2, 2.

Silber, John. 2001. Funding academic science in an age of earmarks. Remarks made at the 26th Annual AAAS Colloquium on Science and Technology Policy, plenary session panel, Washington, DC, May 3–4.

Tresaugue, Matthew. 2001a. Building won't be named for Brown. *The Press Enterprise* (Riverside, CA), January 9, B5.

————. 2001b. College taps U.S. largess: Cal State San Bernardino hops on the grant train. *The Press Enterprise* (Riverside, CA), January 3, A1.

10

The Public-Private Divide in Genomics

Rebecca Eisenberg

In this chapter, I will be looking at the relationship between the public- and private-sector research efforts in genomics, including specifically the publication by the publicly funded Human Genome Project in *Nature*,[1] and the publication by the private, for-profit firm Celera Genomics in *Science*,[2] of their respective versions on the substantial completion of the human genome sequencing project.

The government is a huge sponsor of biomedical research, and although in the past decade the private sector has overtaken government spending in this area, the government is still close behind. It is not really clear what we are to make of these private dollar figures, and it is not clear how much is really properly coded as research or as marketing. There is a lot of government sponsorship, as well as private sponsorship, of research. Both are important, and they are pervasively intertwined. In fact, a lot of this private-sector research is drawing on taxpayer-funded research. The Bayh-Dole Act,[3] with the Stevenson-Wydler Act,[4] and the Federal Technology Transfer Act[5] are a series of related statutes enacted throughout the 1980s and 1990s deliberately promoting private-sector development of advances emerging from government-sponsored research under patents, exclusive licenses, and for profit. Yet we think there is something shady or underhanded taking place when these other firms make a profit.

Look at the story about Yale University's license to a drug company concerning an AIDS drug that they had developed under government funding.[6] On the other coast there is the controversy at Caltech over patents on DNA-sequencing machines.[7] These are discoveries that, pursuant to the scheme of the Bayh-Dole Act, have been licensed to the private sector for development for profit on an exclusive basis. And when these prove to be successful

innovations and profitable innovations, it is the universities that get scrutinized. People want to know about those taxpayer dollars and why these taxpayer-funded inventions are not more broadly and cheaply available.

Using recent events in genomics as a focal point for reexamining the relationship between public and private research, I would like to discuss some of
these issues. I am particularly interested in two public policy mechanisms—
public research funding and intellectual property rights—and how, in fact,
these two mechanisms interact, determining the boundaries between public
and private research. We might think of intellectual property rights and public funding as alternative mechanisms for promoting research in the public
and private spheres, but in contemporary society, they are interrelated in
more complex ways.

Contemporary genomics research is an interesting place to examine this
issue because it presents an interesting challenge to standard accounts concerning how to draw the line between public and private science. There are a
number of reasons for this. The first is that it is really hard to distinguish basic
science from applied technology in this setting. In fact, I think it is impossible. Fundamental scientific breakthroughs in this area often have conspicuous commercial potential. There is substantial overlap in the projects that are
pursued in traditional public research settings, such as university and government labs, and traditional private research settings like laboratories in private
firms. In both public and private settings, there is considerable interest in
using intellectual property rights to appropriate the value of research results.
This is partially due to the Bayh-Dole Act and subsequent related statutes that
encourage institutions doing research with federal funds to pursue patents on
the resulting discoveries, and partly due to the increasing availability of private funds for research in genomics, including university-based research.

A related phenomenon that I find particularly intriguing is that, in this setting, we see outright competition between the public and private sectors in
the pursuit of research goals. While this remains somewhat unusual, we
have seen it on a recurring basis in genomics. This is most dramatically
exemplified by the recent and ongoing race to complete the DNA sequence
of the human genome between the publicly funded Human Genome Project and Celera.

My own attention, however, had been previously focused on this issue by a
couple of back-to-back phone calls that I received in 1998, before Celera even
existed. The first phone call came from a congressional staffer working for
Senator Pete V. Domenici (R-NM). This staffer told me that the senator had
received complaints from a biotechnology firm that the NIH is hostile to
patents and how this is bad for industry. The staffer also wondered why the
federal government was funding research through the Human Genome Project that was competing with work being done by U.S. firms. "Isn't that bad
for the U.S. economy?" she asked.

The second phone call I received was from Francis Collins, the director of the National Human Genome Research Institute at the NIH. He was concerned about the patent implications of private efforts to identify an important new set of genetic markers called "SNPs" (pronounced "snips"), an acronym for single nucleotide polymorphisms, single points of variance in DNA across individuals. These single points of variance are more abundant and easier to use than the markers that formed the basis of the maps that people were then using of the genome, and it was an important resource for future research. He was wondering if these private firms could patent these SNPs, and what the implications of that would be. He was asking about what he should be doing to ensure the availability of SNPs for researchers working in the Human Genome Project planning to create a set of SNPs for the public domain.

Now the juxtaposition of these phone calls was kind of interesting: both of these stories suggested a deeply counterintuitive model of the relationship between public and private efforts in genomics, in which they are competing rather than complementary. And outright competition and rivalry between public and private sectors to fund research makes little sense in standard accounts of the justification for public funding for research. A scientist's view of that relationship would probably look something like Vannevar Bush's account in his post–World War II pitch for continued government funding of basic science in *Science: The Endless Frontier*.[8] This account pictures publicly funded research as promoting activities that are really quite distinct from the type of research that private firms are likely to pursue with their own funds. Bush, in his classic argument for continued funding of research after the conclusion of World War II, called for government funding of basic research to compensate for inadequate commercial incentives to invest in the pursuit of fundamental knowledge. Bush writes,

> Industry will fully rise to the challenge of applying new knowledge to new problems, the commercial incentive can be relied upon for that. But basic research is essentially noncommercial in nature. It will not receive the attention it requires if left to industry.[9]

This suggests a kind of taxonomy of research in which basic science and applied technology are distinct enterprises with distinct goals. Drawing the boundaries between public and private research is a relatively simple task in this stylized world. Basic scientific research is not pursued for profit, is not profitable, and is going to be publicly funded, or otherwise it will not happen, while applied technology research will be privately funded. But that does not seem very helpful in figuring out what is going on in genomics.

Economists offer a slightly different perspective of why government funding is needed for research. According to the economists, the case for public funding of scientific research rests on a disparity between the high social value of research and the low private returns from investments in the pursuit of fundamental knowledge. Private returns are low because of the difficulty of appropriating the value of fundamental research results through patents, and because of the difficulty of foreseeing which firms will profit from research that yields knowledge having unpredictable practical applications. Public funding of fundamental research in institutions that will disseminate their research results broadly solves this problem. The government pays for socially useful research without worrying about having to exclude free riders from enjoying its benefits. You can simply make the results freely available, allowing benefits to fall where they might. In this account, public funding is unlikely to duplicate private investments, although it might yield benefits to private firms. Public funding in this account is complementing and supplementing private funding when private incentives are likely to be inadequate to meet the public interest in scientific progress.

So, keeping in mind my two phone calls concerning why we have public funding of research, the questions are as follows: why should the government pay for research that the private sector is willing to pay for? And the underlying question is really the question that was posed to Senator Domenici: why is the public Human Genome Project competing with the private sector?

There is another account of the point of federal funding of scientific research that I want to put on the table, and this account is more persuasive to politicians than to scientists or economists. But because politicians are the ones who decide how much money to allocate to federally funded research, the perspective that university knowledge will promote economic activity is an important point of view to understand. The political spin on this is that public funding of research promotes U.S. industry and American competitiveness in the world economy by promoting technological innovation by U.S. firms. This is the jingoistic logic of the Bayh-Dole Act and the Stevenson-Wydler Act, both of which direct universities to give preference to U.S. industry in extending licenses to preserve the spillovers generated by U.S. university research for U.S. firms.

This argument is different from both the Vannevar Bush and market failure arguments I just outlined; it has different premises and different implications. The standard accounts presume that research results cannot be appropriated as private intellectual property. That is the basis for the worry about market failure and for the worry that we cannot count on private firms to do this research without government funding.

By contrast, intellectual property plays a central role in the competitiveness account, which argues for facilitating private appropriation of government-funded research results as intellectual property, both to motivate additional private investments and to protect U.S. firms from foreign competition. This account urges the government to pay for research and then hand it off to the private sector. This is the procompetitiveness thinking that underlies the Bayh-Dole Act and other federal programs that focus on harnessing the benefits of U.S. research investments for U.S. firms. In this account, publicly funded research is pictured as an advanced scouting mission for the benefit of shortsighted or risk-adverse U.S. firms. It identifies new profit opportunities that the firms might not find on their own. Although the logic is different from the traditional market failure rationale, the presumption is that public funding is compensating for insufficient motivation in the private sector to pay for research.

A primary concern of political opponents of this competitiveness way of thinking about public research funding is that public funding for competitiveness purposes might displace private investments that firms would otherwise make on their own, but that they would be happy to have the government make for them. And this account actually suggests, I think, another way in which the public-private research rivalry is surprising. Earlier I asked, "Why would the government want to pay for research that the private sector is willing to pay for on its own?" There is a corollary to that question that is equally puzzling, "Why should private firms object if the government is willing to pay for research that they would otherwise have to pay for on their own?"

Juxtaposing these two questions, asking why should the government pay for research that the private sector is willing to pay for, and why should private firms object if the government is willing to do that, it seems that intellectual property rights are of paramount concern on both sides. The private sector might want to pursue research with its own funds, even though the government is willing to pay for it, in order to establish priority for intellectual property purposes. Conversely, the government might want to pursue research that the private sector is willing to pay for in order to defeat, or forestall, foreseeable intellectual property rights claims by putting the information in the public domain. Indeed, that is, in fact, the argument that the public sponsors of the Human Genome Project made in trying to distinguish what they were doing from Celera's work. They were trying to argue for a continued, and indeed accelerated, funding of their research, notwithstanding that there was a private firm that was doing something pretty similar. They were basically saying to Congress, "We need more funding. If we let them get ahead of us in this effort, they'll lock up the genome as private intellectual property, and that would be bad, so give us more money and we'll step on the accelerator and make the genome freely available to everyone."

Intellectual property rights are not all that are at stake in the public-private rivalry. There are other recurring features of public-private competition in genomics, including conspicuous scientific rivalry, different scientific approaches, and this kind of "tortoise-hare" contest over the relative merits of speed versus thoroughness. I think these are interesting and important, and I do not dismiss the unmistakable scientific rivalry that is going on here as a matter of individual personalities caught up in a battle of egos. What is most interesting to me, however, is the intellectual property rights dimension, which I think is genuinely an important driver from both sides. It is also interesting that this was the argument placed front and center in the public forums, including in Congress, while the scientific rivalry aspect has been downplayed and even denied, however implausibly. When Francis Collins sought accelerated funding for the Human Genome Project from Congress in the face of Celera's entry into the field, he did not say, "Give me more money so that my people can claim the glory of completing the first sequence of the human genome for the government." He cited the importance of having sequence information made promptly and freely available in the public domain. This was the same story that the Human Genome Project gave to the media; they said it was important for the public project to keep going because it was important to get the information into the public domain.

Celera's response was interesting. They responded by saying, "We will do that too. We promise to make the raw sequence data freely available." Although their articulation was always carefully hedged, and they have in fact backpedaled considerably from broader interpretations of their care-fully hedged language, they are indeed making some version of the sequence somewhat available. They are not charging any money to look at the sequence—at least if you only want to look at a small portion at a time—but they are controlling access. You have to sign noncompetition agreements to download the whole thing. Nonetheless, it is interesting that their response was to say, "Oh, public availability? We can do that."

So what are we to make of all this? How should Congress be approaching this? Does it make sense for the government to pay for research that competes with the private sector for the purpose of forestalling or getting around intel-lectual property rights? I think this approach raises a number of concerns. Funding research that is competitive with research being done in the private sector for the purpose of forestalling intellectual property claims raises a number of problems.

First of all, it is potentially a waste of resources. Arguably it is a waste of resources to invest in something that is already happening. Another problem is that it simply may not work. Public-sector research initiatives that are responding to private-sector work in progress may come too late—it may be ineffective at forestalling priority if the private research is already under way. It takes time to motivate government research funding. You have to first get legislative appropriation to cover the cost, then you need to solicit proposals, and they need to go through peer review, which is a very slow mechanism for getting out there and getting research done and deposited in the public domain. By contrast, the private sector can allocate funds much more quickly if they see a profit opportunity.

Finally, there are legal obstacles to getting publicly funded research into the public domain. The Bayh-Dole Act presumes that intellectual property rights are positive things, that they promote innovation, and that they are good for U.S. firms. Bayh-Dole reflects a concern that government funding agencies such as NIH may be too inclined to discourage patenting when patenting is a positive thing. There also may be resistance from grantees who want patents and who don't want to give up their right to pursue patents in the interest of making information promptly available in the public domain.

There are many reasons that this is a policy we should think carefully about. What can be said for it, if anything? I think there actually is a lot that can be said for it. I think you have to start with the premise that, at least sometimes, information will be more widely disseminated and more widely used, yielding more knowledge and more useful results, if it is freely available rather than if it is rationed. This is plausible but very contrary to the logic of the Bayh-Dole Act, which presumes that the public domain is bad for the dissemination of new information. The logic underlying the Bayh-Dole Act pictures the public domain as a treacherous quicksand pit in which new discoveries sink into oblivion because the private sector will not touch them since they will be unable to get the patents.

The intuition that favors free availability of information is what economists call "nonrival"—that is, information that many people can use at the same time without interfering with each other's work. This is unlike a car, for instance, which, lent to you, I cannot use until you return it. Some people say that information is a public good, or that scientific knowledge is a public good, but that is not quite right. Richard Nelson defines information as a "latent public good."[10]

This means that, in fact, information is nonrival in the sense that many people can use it at the same time, but this does not necessarily mean that information is nonexcludable. We can make information excludable through intellectual property rights as a matter of social policy if we decide to do that. We sometimes decide to do that in order to enhance private incentives to develop that information. But we do not have to do that, and maybe we do not need the private incentives when we have public funding.

The Bayh-Dole Act creates a strong presumption in favor of making research results excludable. It presumes that, even when you have public funding, you still need incentives from exclusivity to promote commercial development in all but extraordinary circumstances, and it therefore promotes patenting. This presumption now has the force of law, but it is questionable in many cases. It is particularly questionable in an era when open sourcing seems to have some conspicuous successes in the area of information technology, and many private firms are exploring business models that are built on free disclosure of information over the Internet.

However, I do not think we should let go of the propatent presumption lightly. The impact of patents varies greatly across fields, and in some fields they do very little work in promoting innovation. In other fields, they do a great deal of work in promoting innovation. Biomedical research is on the "patents are important" end of that spectrum because it depends heavily on the private sector, which depends heavily on patents. The Bayh-Dole Act sets up roadblocks to putting information in the public domain in recognition of that.

But even the Bayh-Dole Act recognizes that there may be extraordinary circumstances that can justify withholding patents in a particular area. Maybe another way of framing our policy question is, "When, if ever, should a government funding agency, such as the NIH, try to step over the roadblocks that the Bayh-Dole Act puts in their path to fund research for the public domain that competes with private, proprietary research?" When should we believe that the usual logic in favor of private appropriation does not apply to a particular research project or goal? When should we expect that particular research results will have greater social value in the public domain than if they are privately appropriated as intellectual property? Social value includes value to researchers and institutions in both the public and private sectors. And you do sometimes see private firms, even ones engaged in biomedical research, figuring that for certain types of information they are going to be better off putting that information in the public domain.

We have seen this supposition a couple of times in the area of genomics. Merck, in the mid-1990s, decided to fund a public domain set of gene fragments through the Merck Gene Index, or the Merck Genome Index.[11] They decided to fund a university-based effort to sequence portions of genes that they would make freely available in the public domain. We have seen it more recently with the SNP Consortium of major pharmaceutical firms that are trying to identify these common points of variance in the human genome and make those freely available in the public domain.[12] So it is not just commercial naïveté that might suggest that certain types of information could be more effectively utilized in the public domain. When private firms that rely heavily on patents decide to pay for efforts to put information in the public domain, this is presumably a pretty good sign that this is the kind of resource for which the usual propatent presumptions might not apply. However, we do not necessarily want to wait for that sign before we decide that the government should pay for research and put it in the public domain. This is because if there are private firms putting that information into the public domain, we do not need the government to do the same. I think these private-sector efforts to put information in the public domain are interesting and may be a key to elucidating our fundamental question here about whether information is going to be more valuable if it is disseminated in the public domain.

It is useful to consider what motivates different institutions in the public and private sector to put genomic information in the public domain as a way of understanding the value of a public domain in this context. What motivates these different players to put information into the public domain? What are they trying to accomplish? One thing they are plainly trying to accomplish is the scientific recognition and credibility that comes with publishing research results and making them accessible. Publication of new research results to the scientific community confirms that researchers have achieved what they claim to have achieved, and it triggers scientific recognition. At the same time, public disclosure subjects research results to scrutiny in the larger scientific community, and that exposes errors and promotes confidence in the validity of the results. These considerations are particularly important for controversial research claims and for research under conditions of rivalry, when skeptics or rivals are likely to contest whether others have accomplished what they say they have done.

This is something that is salient in the competitive genome efforts and may explain why Celera was so eager to publish their paper in *Science*. The perennial rivalry to establish priority of scientific discovery was aggravated in this particular context by public statements from each side that the other was pursuing an inadequate scientific strategy that was not going to allow them to complete the job that they had said they were going to complete. In order to show that results are complete, one needs to submit the information to the skeptical scrutiny of peer review.

I think a related point that has to be made is this: who cares about scientific recognition? You care about scientific recognition if you need access to top scientific talent. The sponsors cannot ignore the motivations of the scientists who are working for them, and scientists are motivated by recognition. Recognition depends on publication, and I think this explains why Celera, a firm hoping to profit from selling access to proprietary databases of DNA sequence information, agreed to make the raw sequence of the human genome somewhat available free of charge. Without that sort of public disclosure, their claims to priority in completing this scientific accomplishment would be impossible for the scientific community to assess.

Human Genome Sciences has often claimed to have sequenced X number of genes or X percent of the genes, the expressed genes in the human organism; they make these claims by press release, and the scientific community then meets them with skepticism.[13] These claims are understood as advertising, not as science. Proprietary databases of sequence information are subject to some scrutiny. They are available to private subscribers who pay for licensed access to the data, but these researchers may not be the most credible judges about claims of completeness or accuracy of the information that is in the databases. When researchers make their results freely available in the public domain, those results become available to hostile rivals, as well as to sympathetic clients and collaborators, making the claimed accomplishments more credible. Because of the public disclosure, the scientific community was more willing to give greater credence early on to the claimed accomplishments of the public-sector DNA-sequencing efforts than to the claimed accomplishments of their counterparts in the private sector, which were not backed up by data release. Publication by press release might be enough to bring in corporate subscribers and shareholders, but not enough to bring in a Nobel Prize. Again, one must consider the motivations of the scientists that you want to have working for you.

Apart from scientific recognition, credibility, and access to top scientific talent that is motivated by recognition, scientific validation may also have commercial value. The credibility of your research claims may be important commercial information. I think this is something that is visible in the SNP Consortium, in their determination to make their polymorphisms, their markers, freely available in the public domain.[14]

These firms in the SNP Consortium hope to use these polymorphisms as "pharmaco-genomic" markers that will predict drug response in individual patients. They hope that this will help them to gain regulatory approval to sell drugs that are safe and effective in some patients but dangerous or ineffective in others. If you cannot distinguish the people for whom the drug is safe and effective from the people in whom the drug is going to cause acute liver failure and death, it is not a safe or effective drug. You cannot bring the drug to market. But if you have a diagnostic test that you can administer in advance that will identify who will benefit and who will suffer, then you may have a safe and effective product.

They are hoping that, if the patient who stands to benefit from a drug may be distinguished through the use of these diagnostic markers to predict the drug response, the FDA might be persuaded to approve the drug for sale to the genetically screened patients, even if it is not safe and effective for the entire universe of people suffering from the disease. They also expect that regulatory approval for these products will turn on the scientific credibility of the tests. The scientific credibility of the tests, in turn, they think, will be easier to establish if the test uses markers that are in the public domain. Why? Because those markers have more scientific credibility since they were and are subject to validation and challenge in the public domain.

Another reason why you might want to make genomic information freely available in the public domain, something that may be motivating both public- and private-sector institutions to disclose that information, is to promote widespread dissemination and use apart from the recognition and credibility of research claims. Some research sponsors can plausibly claim to favor free disclosure of DNA sequence information in publicly accessible databases as a means of promoting dissemination and utilization of research results. The public domain is obviously a winner for this purpose: it is accessible, easy, and there is no transaction cost. Cheapness is particularly important for promoting access by impecunious users, such as academic researchers—people who do not have millions of dollars to spend on a database access agreement. More people will have access to discoveries that are made freely available, and more people will have an opportunity to use those discoveries and build upon them in future research.

However, this is a concern for the pharmaceutical industry as well, which is why one can see private industry sponsoring some of these public domain genomics efforts. The pharmaceutical firms that invest in putting DNA sequence information and SNPs into the public domain may figure that they are going to save on costs—future costs of access to proprietary databases—if there is something that they can use that is in the public domain.

Moreover, private firms may benefit from facilitating academic research. This is an argument that Merck made quite explicitly when they decided, some years ago, to fund the Merck Genome Initiative to put information on expressed genes in the public domain. Merck said that it thought the company would profit sooner and more quickly if this information were made freely available to academic scientists who would expand upon it and learn more about it and bring more opportunities for drug development into view.[15] I think that this is a plausible claim. Pharmaceutical firms profit from developing and selling drugs, and they might believe that they will earn more profits at an earlier date from drug developments if they can accelerate progress in fundamental biological research that is going to bring more drug targets into view sooner and identify more disease association sooner. Rather than trying to do all this fundamental research themselves, which is an expensive job at which they have no comparative advantage, they would be happy to have it done by universities working with public funds, and they want to be sure that the universities are in a position to do that.

So, to the extent that free access throughout the scientific community facilitates this sort of premarket research in universities, pharmaceutical firms may find that their interests are aligned with the interests of public research sponsors in promoting free disclosure of DNA sequence information in the public domain.

Finally, another interest that is served by putting information in the public domain that cannot be dismissed is to defeat patents claims by creating prior art that will limit what other institutions may patent in the future, and this in part because it has been explicitly claimed by a number of the institutions that are sponsoring these public domain efforts. In order to get a patent on something, it has to be new and nonobvious.[16] It cannot be something that was disclosed in the prior literature in the field or made obvious by prior disclosures.[17] So public disclosure of DNA sequence information expands the universe of prior art, thereby limiting what may be patented in the future. The Human Genome Project has been relatively up front about the fact that they require deposit of DNA sequence information generated with Human Genome Project funds within twenty-four hours in the publicly accessible database, GenBank. It is also pretty clear that one goal of this requirement is to prevent the patenting of as much sequence information as possible. This accelerated timetable makes it difficult for grantees to get patent applications on file prior to public disclosure, in addition to sorting through the newly identified sequence information to figure out if it includes any sequences that are worth patenting.

These prompt-disclosure rules also lead to the prompt creation of prior art that could potentially defeat patent claims based on similar DNA-sequencing efforts in the private sector. No one can get a patent on something that was already publicly disclosed before the patent claimant discovered it. So a research sponsor might want to create patent-defeating prior art for all of the reasons it wants to put information in the public domain in the first place. The SNP Consortium openly acknowledges this goal on their Web page, where they state, "The SNP Consortium was created precisely to produce a high-quality SNP map that will be publicly available and freely accessible."[18] It is a very candid admission of what is motivating them to put information in the public domain.

A final argument for putting this information in the public domain that cuts across the interests of the public and private sectors is avoiding fragmentation of intellectual property rights in the sort of resources that are best used in large aggregations. This is a particularly salient concern for SNPs—for markers on a map. You do not want a map with one or a small number of markers. You want as many markers as possible in order to have the most complete map. It is better to have everyone working off of the same complete map rather than to have numerous less complete proprietary maps; however, gathering together licenses from multiple patent holders could be a problem.

There are also a number of other reasons why public and private firms might think it is valuable to have genomic information placed in the public domain: because people will get more value out of it; the information will be more effectively, cheaply, and widely utilized if it is in the public domain than if it is privately appropriated; and so on.

Even so, if we accept that genomic information may be more effectively utilized in the public domain than if it is all locked up in proprietary databases, does it necessarily follow that government should be competing with the private sector to discover it? Or is that the wrong response? Maybe rather than drowning the patent race with public funding, we should instead be focusing on fixing the patent system so that none of this stuff gets patented. Maybe we are playing around with the wrong lever if we view public funding and intellectual property rights as alternative public policy mechanisms for mediating this public-private divide. Even if this is what a unitary government decision maker would do, it is not what we have.

We have a patent system on one hand and a research funding system on the other hand. These are administered by different people with different concerns and different types of institutional competence. Within these real-world constraints, there are reasons to think that it might make more sense for the NIH to try to fix the problem with grants rather than trying to get the Patent and Trademark Office (PTO) to fix the problem by fine-tuning the rules of what can get patented.

First of all, it is important to recognize that we are not simply looking at patent rights here. Withholding patents might not do the trick because there are other intellectual property strategies that could have the same effect. Private firms could still keep genomic information in proprietary databases and ration access to them, which would have much of the same effects, although competing databases could arise if there are no patent rights in play, and that would help lower access costs.

Setting that aside, the patent system is less responsive to emerging policy considerations in specific fields than the research funding system for a variety of reasons that may make it much easier to fix this problem at the level of funding decisions. It takes years for the Patent and Trademark Office and the courts to determine what is patentable when the patent system enters a new field. The courts are now considering issues raised by discoveries made twenty years ago. It usually takes many years to change patent statutes if you do not like the law once the courts have clarified it. The patent system is singularly reluctant to address specific policy issues raised by specific technologies. We have a unitary patent system that purports to apply the same set of rules across all fields of technology. And you can see how uncomfortable the patent office is with the role of addressing different concerns arising in specific fields. For example, they have hearings on the utility requirement, where all the testimony is about DNA sequences, and yet they adopt guidelines that apply across all fields. They have a hard time fine-tuning their rules for problems arising in particular fields.

Moreover, the funding system can respond to emerging policy issues in months rather than years. It is slow; they may not be as fast as private firms in deciding how they are going to spend their own research dollars, but it is still much faster than the kind of response one would get out of the patent system. The funding system is focused inherently on emerging questions and future needs, while the patent system is backwardly focused on prior claims. The funding system uses peer review to give access to high-level, up-to-date input from active practitioners in the field.

By contrast, the patent system uses examiners who, although technically trained, are not active practitioners in their scientific and technical fields at all. So if the reason for ensuring free access is to meet the needs of the scientific community, research funding agencies such as NIH are going to have much better information about what those needs are. The PTO does not really have that kind of input. They can hold hearings from time to time, but basically, in deciding on particular issues that are presented to them, they are only talking to the inventor, not to the broader public or getting that kind of broad input.

On the other hand, the competing argument is that the NIH does not really know anything about patents. Agencies tend to overpredict problems with the patent system, which is why the Bayh-Dole Act was passed in the first place, in that agencies have been too reluctant to permit people to patent things where patents were appropriate to motivate private development. There is another problem with trying to fix the patent system rather than trying to address the problem at the level of allocation of federal research funding dollars: fixing the patent system by reducing what can be patented may destroy incentives that are important to private firms, although unnecessary perhaps for government-sponsored research. And I think that is a very dangerous path to follow. Having a mixture of public and private research efforts may have a lot of value for science. Even in efforts to discover fundamental information that is important to the infrastructure of science, private firms may foresee possibilities that governments, or any unitary decision maker, including a private monopolist, overlooks. Patents may be necessary to motivate private firms to keep second-guessing the way the government is spending its research dollars.

In fact, this is exemplified in the case of genomics by Craig Venter, the founder of Celera. The NIH repeatedly passed up Craig Venter's ideas when he sought government funding, good ideas that should have been funded. This was true of his insight to sequence a portion of genes and create a catalog of Expressed Sequence Tags (ESTs) that basically get a snapshot picture of a portion of each gene. This turned out to be a tremendous resource for discovery that the NIH decided they did not want to invest any resources in funding, so he moved it out into the private sector. More recently, his idea of using whole genome shotgun sequencing to complete the sequence of the human genome looked like a crazy idea to the public sponsors of the Human Genome Project and the peer reviewers they brought in to help them evaluate it, but it looked like a good idea to private sponsors: lo and behold, it was a good idea.[19] So you may lose out on some valuable research initiatives that the scientific establishment passes up but that a private firm might be willing to pursue if it has enough incentive to do it. I think we may need to preserve patents to get the benefits of discoveries that the private sector is willing to go for and that the government might overlook, reject, or dismiss.

My bottom line here is a fairly modest one at this point. I think that the interest in widespread access to information may sometimes justify spending public funds on acquiring information, even though private firms motivated by patents are willing to pay for the same thing. In most situations, if you have private firms that are already doing that research and are already willing to fund similar research, then this is probably pretty good prima facie evidence that public funds would be better spent elsewhere. However, this is not always the case. Even when patents are available in an area, and where they are sufficiently valuable to motivate firms to invest in research, patents on fundamental knowledge might not be worth acquiring if they harm future research—and this is a legitimate concern, I think, of public research sponsors. You do not want to let the antipatent reflex get too strong or dominate the setting of research priorities. I think funding agencies are better equipped to find the most promising research opportunities—much better at it than they are at foreseeing patent problems. Nonetheless, I think the public interest may sometimes, and perhaps even often, be better served by widespread availability of new knowledge. The public research sponsors have reason to be skeptical of the strong Bayh-Dole Act presumption to the contrary, that private appropriation is always and inevitably a good thing.

Following are Professor Eisenberg's responses to questions posed during a discussion of her paper at the University of Illinois Center for Advanced Study.

Q: You suggested that funding agencies could drown out the patent process by putting information in the public domain, and that one way this could happen is by demanding that the information is released very rapidly, like in the Human Genome Project. How else could they do that in an effective way?

A: *There is a provision in the Bayh-Dole Act that explicitly addresses this.[20] I mentioned earlier exceptional circumstances: the agency can make a determination that these are exceptional circumstances where the usual propatent rules do not apply, and therefore, for this particular funding initiative, the agency is not going to allow grantees to patent the information. That would be the straightforward way that the statute contemplates for the NIH to do that. But that could be all tied up in grantees promptly saying, "We don't want to go that way. We disagree. We don't think these are exceptional circumstances; we think these are ordinary circumstances and that we should be allowed to hold onto our patent rights." In order to get out there quickly, they were trying to do an end run around that mechanism by being silent on the issue of patents and just saying, "We require you to deposit your sequence information promptly in GenBank," thinking that they were thwarting those efforts.*

More recently they have gotten a little bit more outspoken about this. They have gotten a little bit bolder about their agenda of preventing people from patenting things. When they moved to the whole genome-sequencing phase of the Human Genome Project, they asked their grantees to address the question, "What are your plans for making this information promptly available? We don't think that large chunks of the genome are appropriate candidates for patent protection, and we will consider that in evaluating proposals." So, they became bolder about their antipatent stance, but they never went through the procedure contemplated by the Bayh-Dole Act of making a finding of exceptional circumstances.

Q: I wonder if you could respond to a couple of other possible policy instruments in areas where, using all the factors you talked about, the government decides that research ought to be in the public domain. One is the use of eminent domain to just let the private party develop whatever intellectual property they want and then condemn it, and then it is in the public domain because you have bought it. Is this economically or legally feasible? Second, you mentioned something about proprietary databases, which suggests that some research was using trade secret or forms of intellectual property other than patent law, requiring you to put all the information in the public domain. Would the other alternative be requiring that patents are the only acceptable way of intellectual property protection, so that at least the data would all be out there in the public record in order to receive the benefit of the patent?

A: *Addressing the issue of the eminent domain first—there is authority for the government to use any patented invention subject to a damage remedy in the court of claims for the government's own activities. That is an eminent domain statute that is on the books. In the case of government-sponsored research discoveries, through Bayh-Dole Act patented inventions, the government has a retained license to use the invention for government purposes and to authorize others to use the invention for government purposes. There is a question as to how broadly that language should be construed. The NIH has been very conservative in its position on that, but they could take a more aggressive position and say, "We think that our grantees have a retained right to use Bayh-Dole discoveries in their NIH funded research." Some agencies have taken that position. The NIH, for a variety of reasons, has not. Eminent domain is certainly a viable strategy. It could be fairly expensive, particularly if the patent system is giving very valuable patents in exchange for trivial advances. When you see a lot of people falling all over themselves competing to do the same thing, as in what is going on in this genomics area, that is a good clue that you are overmotivating investment in that particular type of research. The fact that we are seeing all of this racing, I think, indicates that eminent domain would be an expensive strategy, because the patent rights are probably a lot more valuable than the cost to the government of just doing it itself and making it freely available.*

As for forbidding trade secrets, the Bayh-Dole Act is really about patents rather than trade secrets. In effect, the NIH does require disclosure of research results, but it is hard to imagine eliminating trade secrecy entirely in a system that contemplates that people are going to pursue patent protection, because disclosure limits what you can patent. You at least have to countenance some interim secrecy while people get patent applications on file. I suppose you could be a little bit more explicit about that. I don't know that you could go so far as to forbid trade secrecy. But you could work with your disclosure policies—these NIH genome disclosure rules are one extreme example, but there are other ways that would give people some latitude to pursue patents via interim secrecy while still limiting long-term secrecy.

Q: Directing your attention away from the legalistic aspects toward the broader scientific issues, I think by couching the issue of whether and how the government should support basic research or research in universities in terms of the recent genome issues and biological sciences you are perhaps giving a bit of a distorted view for a few reasons. First, I think the assumption made in the whole genome project is that the transition between basic research and applications that bring in real dollars is relatively short, and that is in large part based on the last ten or fifteen years of developments in the semiconductor industry, where there has been a short transition between basic developments and applications that make money. But if you take a broader view, in many areas of science there is a twenty- to thirty-year transition between the basic science and the actual project. Another reason why the companies have given up on basic science research in, say, condensed metaphysics, with the exception of two companies—IBM and Lucent—is that it is just not profitable for them to invest in basic research if there is that twenty- to thirty-year transition to a product, whereas the government is almost required in that case to fund the research. I wonder if you can comment on that in a broader sense?

A: *I was trying to look at the harder case of whether there ever is a reason for the government to invest in research when industry is funding the same research with the same goals. But I totally agree with you that, in many fields, you cannot rely on the private sector to invest in new knowledge that will ultimately have payoffs for them because their timeline is too short or because they just cannot persuade investors to give them the funds that they need in order to survive for twenty to thirty years. Then it is an easy, clear case for government funding of research.*

Q: If you look at the Commerce Department website, it says, "Copyright claimed everywhere in the world except the United States." Yet the NIH "publication in twenty-four hours" rule might result in patents only in the United States and nowhere else in the world. Is public domain an evil thing that we should inflict on the rest of the world, or is it a good thing that we should keep here and inflict these evil patents on the rest of the world?

A: *I think that the presumption in the system is that patents are good things, but yet maybe the antipatent folks presumably do not think that. As for the specific setting of genomics, I think the assumption is that U.S. patent law is what really matters. First of all, U.S. law has been more generous toward what can be patented in this area. The European patent convention has more restrictive and ambiguous rules. They have been a little slower to face some of the fine-grained questions, but U.S. patents often arise first, and then there is a lot of discussion in Europe about whether the allowance of patents on human DNA sequences is going to affect public order or morality. The combination of the United States being such an important market and U.S. law being basically more hospitable toward protection of biotechnology-related advances has made people more focused on the U.S. patent system.*

As a practical matter, if something is patent free, particularly if you are talking about something that will be an input into future research, the consequence of the United States' generosity with our patent protection might be that a lot of research is going to move offshore. Some people from pharmaceutical firms have suggested to me that if they see too much of a proliferation of patents on drug targets in the United States, they will just move their drug-screening operations offshore and conduct them on a large ship in the middle of the ocean if necessary. Not everybody is free to do that, but you could imagine if we are at the point where patents are creating obstacles to further research on balance, then one solution is to locate your biomedical research and product development facilities elsewhere where you do not have to worry about these patents.

Q: I think all of us who took high school chemistry have seen Mendeleyev's periodic table of the elements. When it came up, it was incomplete. We have been working to fill it. I don't know whether all the holes have been filled in the past 130 years, but I think there is an interesting parallel to genomics. Can anyone imagine what would happen if somebody tried to commercialize Mendeleyev's periodic table of the elements 130 years ago?

A: *Presumably, at worst, we would be seventeen years behind the current state of chemistry research. Patents do expire; a patent is not forever. But yes, the metaphor is a nice one, and one some scientists offer. The metaphor is that this is the sort of fundamental knowledge that forms the springboard for future research, and it is going to be more accessible and more effectively utilized if freely available to more people. One thing that I want to point out—I haven't been focusing specifically on what it means to patent DNA sequences, but for the most part that means you are patenting molecules. The Patent and Trademark Office and the courts have treated claims to DNA sequences as a claim to a chemical, a long sequence of nucleotides. As a result, it is not the information itself, or the organization of the information, that is the subject of patent claims. The first generation of DNA sequence patents were on genes encoding therapeutic proteins like insulin or growth hormone. The biotechnology companies that developed these products would claim that the gene is an isolated molecule, and recombinant materials that incorporated that gene could be used as production facilities for making the protein itself. That looked sort of like a high-tech equivalent of a patent on a drug. It gave the biotechnology company a monopoly over the only commercially feasible way of producing that product in abundant quantities. When you are talking about not a particular isolated and purified gene, but rather whole genomes of organisms at the scale that this information is coming out today, it looks a lot more like patenting scientific information, and that is why this metaphor to the periodic chart, I think, is such a nice one.*

NOTES

1. See generally in *Nature*, February 15, 2001, 409.
2. See generally in *Science*, February 16, 2001, 291.

3. Act of December 12, 1980, Pub. L. No. 96-517, 94 Stat. 3015-28 (codified as amended at 35 U.S.C. 200-11, 301-7 [2000]).

4. Stevenson-Wydler Technology Innovation Act of 1980, Pub. L. No. 96-480, 94 Stat. 2311-20 (codified as amended at 15 U.S.C. 3701-14 [2000]).

5. Federal Technology Transfer Act of 1986, Pub. L. No. 99-502, 2, 100 Stat. 1785-87 (codified as amended at 15 U.S.C. 3710a[a][1], [b][2]-[3] [2000]).

6. Melody Peterson, "Abbot to Sell Low-Cost AIDS Drugs in Africa," *New York Times*, March 28, 2001, C9.

7. Peter G. Gosselin and Paul Jacobs, "DNA Device's Heredity Scrutinized by U.S.," *The Los Angeles Times*, May 14, 2000, A1.

8. See Vannevar Bush, *Science, the Endless Frontier: A Report to the President on a Program for Postwar Scientific Research* (Washington, DC: U.S. Government Printing Office, 1945).

9. Bush, *Science, the Endless Frontier*.

10. Richard R. Nelson, "What is 'Commercial' and What Is 'Public'? In *Technology and the Wealth of Nations*, ed. Nathan Rosenberg, Ralph Landau, and David C. Mowery (Stanford, CA: Stanford University Press, 1992), 57, 61. Here, they refer to technology as a "latent public good."

11. Eliot Marshall, "The Human Gene Hunt Scales Up, *Science* 274, no. 1456 (1996).

12. Andrew Pollack, "New Venture Aims to Guard Genetic Data," *New York Times*, October 9, 2000, C2.

13. See www.hgsi.com.

14. See supra note 13.

15. Rebecca S. Eisenberg, "Intellectual Property Issues in Genomics," *Trends in Biotechnology* 14 (1996): 302-7.

16. 5 U.S.C. 101 (1994); 35 U.S.C. 103 (1994).

17. 35 U.S.C. 102a (1994).

18. The SNP Consortium Ltd. website, available at: *http://snp.cshl.org/about/faq.shtml*.

19. "Venter Receives Common Wealth Award," *American Scientist* 381 (July 1, 2001): 89.

20. See supra note 3.

Section IV

THE DARK SIDE OF UNIVERSITY-CORPORATE PARTNERSHIPS

Michael Hansen opens this section with a highly critical account of what he views as the tendency of university-corporate relations to skew research agendas in favor of short-term profitable goals and away from longer-term agendas in the public interest that may not involve profit. Toby Miller then describes a corporatized university in which research institutions have become, in his view, "landlords, tax havens, and R & D surrogates rolled into one, with the administrators and fundraisers lording it over the faculty." Maso Miyoshi closes with a lament that "higher education as a whole is currently experiencing a nearly complete loss of its historic purpose," with a growing divide between faculty engaged in commercially valuable pursuits and those who are not threatening the future of research universities as we know them.

11

The Effects of University-Corporate Relations on Biotechnology Research

Michael K. Hansen

I have had the opportunity in my research to focus on nonprofit-sector work on genetic engineering, on pesticides and their alternatives, and, more recently, on bovine spongiform encephalopathy (BSE), aka Mad Cow disease. In these areas, my efforts have been to get U.S. and international agencies to properly regulate these technologies and to require adequate testing. I have come across a myriad of problems in this area due to the increasing relationship between industry and universities. I would like to present an overview of these problems and try to trace them back to their roots.

We can trace many of the problems associated with increasing university-industry connections to two key pieces of legislation: the Bayh-Dole Act and the Technology Transfer Act. The rationale of the Bayh-Dole Act, passed in 1980, was to make research in the universities more available to the public by allowing universities to patent inventions developed using federal grant money from the National Institutes of Health (NIH) and the National Science Foundation (NSF), and, more controversially, to license these inventions to industry in return for royalties.

This was a big step forward, especially looking at the congressional climate post–World War II, which is when the NSF was established. Some of the issues raised then were the same issues facing us now. There was concern in the 1950s that providing federal funds for research would allow university scientists to use money provided by the public to do research for personal gain. The potential for scientists to profit from their work could also skew the kind of research that they choose to do. This directly conflicts with the purpose of public-sector research, which is to provide for the public good. The Bayh-Dole Act permitted universities to patent the results of their federally funded research and allowed them to share some of the profits with the researchers through royalties. As patenting increased, industry funding increased.

The Technology Transfer Act of 1986[1] extended the Bayh-Dole Act by mandating the establishment of Cooperative Research and Development Agreements (CRADAs), which permitted researchers receiving public funds to enter into R & D agreements directly with private companies and to share the profits of their research results.

The fallout from this legislation has been phenomenal. In 1985, the overall amount of corporate funding going into the university system was $850 million; by 1994, it was $4.25 billion—a substantial increase.[2] Curiously, only about half actually went toward research. Much of the money was being allocated for the purchase of endowed chairs and for putting corporate logos on different paraphernalia. For example, K-Mart funded an endowed chair in the management school at West Virginia University.[3] One of the conditions of the endowment is that the person holding the chair must train assistant store managers for K-Mart for at least thirty days during the year. At the University of California at Berkeley, corporate funding has increased, while federal and state funding of science research has decreased. Buildings throughout the Haas School of Business in Berkeley have been plastered with corporate logos. The Dean of the Haas Business School is now officially known as the Bank America Dean of Haas.[4]

Looking at industry funding for academic research between 1980 and 1998, industry funding has increased at a rate of 8.1 percent per year so that in 1998, $1.9 billion was being transferred from corporations to university research, which was eight times the level in 1980.[5] There has been a similar growth in university patents. The number of patents applied for by universities throughout the United States was about 250 per year before 1980; in 1998, there were approximately 4,800 potential patents issued to universities.

Another way that consumerism or commercialization is affecting the university is in the growing individualistic attitude among university researchers; that is, there is a general disregard for the implications of research for the community at large. For my postdoctoral work at the University of Kentucky, we conducted an extensive series of interviews with scientists in both public and private sectors, along with plant breeders and biotech and molecular biologists, examining the general impact of genetic engineering on agricultural research. We chose to focus on two crops, tomatoes and wheat. We saw a significant decline in the number of full-time equivalents for conventional plant breeding due to either the status quo or declining federal funding.[6] Consequently, universities were looking for other sources of funding, and biotechnology was thought to be the answer.

One thing we found while interviewing the molecular biologists was that their ultimate goal was to engineer a plant that could function within the confines of the lab. In fact, a dean at an agriculture school in the West corroborated this sentiment by stating that "one molecular biologist will be able to replace four plant breeders." However, the problem is that this type of genetic engineering cannot be disconnected from its context.

For example, in the early 1980s, scientists were trying to put a modified form of an endotoxin from *Bacillus thuringiensis* (Bt), a genus of soil bacteria, into plants. Some scientists figured if they could insert the bacterial toxin gene into the plant, have the gene turned on, and then successfully grow that plant in the lab, the job would be finished. In other words, if the plant worked in the lab, it would work the same way when planted outdoors. I remember talking to one scientist at a company and pointing out the fact that when a plant undergoes that type of engineering, it needs to be kept in its larger natural context. The scientist, who was a molecular biologist and was not familiar with ecology or population genetics, responded by saying, "That is not a problem. The big problem is getting it to work in the lab." While lab success is, in fact, a critical step, it is still one step in the process of introducing something into the environment, because one needs to understand so many different factors, for example, how all the elements interact with each other, how that bacterium interacts with the environment, and so on. Without taking these other factors into consideration, the consequences of introducing genetically engineered plants into the environment, in the worst-case scenario, could have a catastrophic effect by disrupting the ecosystem.

In a slightly less dramatic example, we have seen universities focus almost exclusively on researching synthetic chemical alternatives for pest control in agriculture during the last fifty years, during which pesticides were the panacea. After p,p'-dichlorodiphenyltrichloroethane (DDT) was first used successfully to stop an outbreak of typhus in Naples, Italy, in 1944–1945 and used effectively in World War II, it was released in the United States for the control of agricultural pests and for the eradication of insect-borne disease.[7] Public health scientists talked about eradicating diseases such as typhus, malaria, and yellow fever. At the time, there was talk of spraying every house in developing countries with DDT because they thought it was a wonderful chemical and all pests would be under control.

However, the results of a 1940s study showed that two trees with scale insects (immobile sucking insects resembling little bumps) could have up to a thousandfold difference in the insect levels, reflecting differing degrees of resistance to the DDT treatment.[8] The methodology involved using one tree as the control and the other as the experimental by spraying one tree and not the other. The results were that the sprayed tree had more than one thousand times the number of these scale insects than the unsprayed tree. This was due to the fact that when you sprayed the DDT to kill that pest, the pest's natural enemies were also eliminated, affecting an entire food chain. The insecticide was disrupting the complex food chain among all the predators that fed on these insects. In spite of some of this early work by insect ecologists, the chemical era was rampant. We are now more aware of the myriad of problems with pesticides, so the next era is going to be genetic engineering.

CASE STUDY: CLASSICAL BIOLOGICAL CONTROL

Before discussing some of the problems with university-industry relationships, I would like to examine the core dilemma of doing research for the public good versus doing research for profit by looking at classic biological control. Classical biological control is the use of natural enemies as pest control. Many of the pests found in agriculture in the United States are not native to the United States but come from elsewhere. They are transported from other countries via crops coming into the United States. It was discovered through classical biological control that within the countries where these pests naturally dwell, the pests would often be scarce because natural enemies control them. The obvious solution, then, was in figuring out what are the most effective natural enemies for certain pests and then transporting these natural predators to the United States, resulting in a "natural" solution for this problem. Classical biological control is for the public good and is something that only public universities and the public sector can provide. This type of research, in some sense, is strictly for public good versus gaining profit, since there is no dependent market being created for a product. This is where research intended for the public good and privately funded research diverge.

The first time that classical biological control was implemented in the United States was in the 1880s. The cottony-cushion scale insect had been accidentally brought into California on acacia trees from Australia in the early 1880s. Within a few years, it threatened to destroy the state's entire citrus industry. The industry initially tried to eradicate the insect by spraying the trees with Paris Green[9] (a dye whose insect-killing properties were discovered in 1868) and lead acetate, with no success. In desperation, the University of California sent Dr. Alfred Koelbe to Australia to look for natural enemies. He found that the cottony-cushion scale was rare in Australia, in large part because it was being controlled by a number of natural enemies, including parasitic wasps and the vedalia beetle (a kind of ladybug). In 1888, Dr. Koelbe sent back specimens of the vedalia beetle, which was released and spread throughout the state as it fed on the cottony-cushion scale. Within a couple of years, the beetle had completely controlled the scale.

In classical biological control, funding goes into understanding the biology of the pest in its original habitat, finding the right natural enemy, testing it before releasing it, and making sure that it fulfills its very specially ordained task (e.g., only attacking the pest insect), all the while being careful not to affect other food chains. And then, once released, if successful, the natural enemy will spread itself via procreation. In a strictly economic sense, the natural enemy is a product that destroys its own market.

Herein lies the dilemma: this is something the private sector can never develop, because money would be invested into coming up with a natural enemy without having a large enough market for the product. After a private developer would make a couple of sales, the product would then spread by itself. The private developer's profit margin in the long run would then be drastically reduced, since if you find the solution through this type of self-controlling and sustaining system, the cost is up front. Looking at the few studies that have been done on the return of these investments, results show that biological controls pay back at a rate of one hundred to two hundred dollars or more for every dollar invested;[10] this compares to pesticides, which pay back at the rate of three to four dollars for every dollar spent on them.[11] The cost of the solution is entirely up front, but the benefit is ongoing: not only the basic value of that crop but the commodity every year from the introduction of this type of solution onward. Successful natural enemies do not need to be continuously released; they reproduce by feeding on the pest.

Classical biological control has worked many times globally since the 1880s. For example, the vedalia beetle has been imported into thirty other countries to control the cottony-cushion scale. In addition, the research stimulated by the success with the cottony-cushion scale has led to hundreds of successful biological control programs for a wide range of pests.[12] In some Asian countries, along with the United States and other countries, the beginning of the twentieth century saw an increased interest in ecology and issues concerning biological control. In fact, this increased interest led to the creation of the Center for Biological Control at the University of California at Berkeley, which went on to become a world-class facility. However, it started to go downhill with the advent of the chemical revolution in the 1950s.

In fact, about five years ago, the world-class facility at the Center for Biological Control at Berkeley was shut down. Half of the entomologists have now left, and molecular biology now dominates. Novartis Pharmaceuticals Corporation negotiated a $25 million contract with the University of California at Berkeley, where it will fund one-third of one department at the university for five years. In exchange, the company will get first rights to negotiate on approximately one-third of all discoveries made in that department. This right is not limited to discoveries resulting from Novartis funding but extends to any discovery made in that department during that five-year period. There is a Faustian "flavor" to this deal, as Novartis gets to take the best one-third of the research, regardless of who funded it, with the university getting the rest.[13]

Many Berkeley graduate students and faculty are up in arms about the Novartis-Berkeley deal because they themselves are concerned about the impact that private funding will have. In his defense, Gordon Rausser, dean of the College of Natural Resources and the person who negotiated the deal, said, "Without modern lab facilities and access to commercially developed proprietary databases, we can neither provide first rate graduate education nor perform fundamental research that is a part of the University's mission."[14] In one sense, Rausser seems to be implying that Berkeley cannot fulfill its mission of providing a first-rate graduate education or performing basic research without this type of private funding.

With all that money coming in, people in California and elsewhere may no longer see the university as a neutral arbiter on issues of genetic engineering because they are increasingly seen as being co-opted by money.

One group this affects is the agricultural industry. The University of California has some very interesting programs, including a sustainable agriculture program that focuses on small farms, organic farms, and farmer-to-farmer education. All these farmers are concerned that the money coming into the university is not serving their interests and that there is little to no research being done currently that benefits them.

Besides the closing of the Center for Biological Control, there has been a general decline in funding for integrated pest management, agricultural ecology, insect ecology, insect taxonomy, and so on, as well as declining federal and state funding for agriculture. Universities and scientists are scrambling to procure funding. It is an anomaly that these sustainable agriculture programs receive little or no funding in light of the fact that study after study shows that these programs work. The problem with sustainable agriculture approaches, such as a crop rotation scheme, use of cover crops, or activities to empower farmers to understand the biology of their agricultural systems in order to manage them in a more sustainable manner, is that there will be a reduction in synthetic chemical use. With fewer or no chemicals to sell, then, there is less money to be made, and the universities are not as interested in that kind of research. This "profit-driven" mentality is disconcerting.

While some of these problems could be addressed by the private-sector research, areas like classical biological control can only be done in the public sector. For example, Don Dahlsten, a former head of Cal's Center for Biological Control, pointed out the huge success in Africa with the cassava mealybug and the cassava green mite. Cassava is actually a root crop consumed by about 250 to 400 million people in Africa and is the main crop that people eat. The cassava mealybug and the cassava green spider mite were accidentally introduced into Africa from South America around 1970 and swept through Africa.[15] In some countries, these pests caused up to 60 percent yield losses in cassava, leading to serious starvation problems.

In response to this crisis, a massive classical biological control project was undertaken in 1980 by the International Institute for Tropical Agriculture (IITA), located in Ibadan, Nigeria, with the help of the University of California at Berkeley. It took them about $5 million to work out the problem. For the first five years, the scientists could not identify any parasitic insects that attacked the cassava mealybug. The reason for their lack of success was that the cassava mealybug they had identified was not the proper species. The taxonomy had yet to be done, which would reveal that there were, in fact, two species of mealybugs. Because of this oversight, the parasitic wasps, brought in from South America to control the cassava mealybugs, were not alleviating the African cassava mealybug problem. Once they identified the proper species, the scientists went back to South America and found the proper natural enemy, a little parasitic wasp called *Epidinocarsis lopezi*, and, more recently, also a spider mite that effectively controls the African cassava mealybug population.

The project then had to develop a methodology to rear huge numbers of parasitic wasps and release them by airplane throughout the entire cassava-growing region.

This biological control project has been considered a huge success throughout Africa. By 1985, the wasp had spread over 570,000 square kilometers (or about 12 percent) of the 4.5 million square kilometers on which cassava is grown. By the early 1990s, the wasps had spread over the entire cassava-growing belt, an area larger than the United States, with the result that the mealybug problem had been effectively eliminated in thirty African countries without the use of pesticides. The estimated benefit of this project to African farmers is about $3 billion. In 1995, Dr. Hans Herren, an IITA entomologist who helped conceive and lead the project since 1979, was awarded the World Food Prize.[16]

Strict classical biological control creates a product that destroys its own market, but it is done in the public's best interest. Such research is the appropriate first line of agricultural research that needs to be focused on at the land-grant universities.

In a slightly different example, also from California, the Sandoz Pharmaceutical Corporation[17] attempted to set up a nonprofit in 1992 to make a deal with the Scripps Research Institute, a large nonprofit research organization.[18] At the time, the deal was controversial, in part because Sandoz wanted to pay $300 million over ten years to the Scripps Research Institute. In return, Scripps was to give Sandoz first option to license virtually all of the institute's biomedical research resulting from their funding, as well as fruitful research funded by the NIH. Sandoz was basically asking for virtually all research that would be produced from the research center.

PROBLEMS WITH PRIVATE FUNDS

With the influx of private funding coming into research universities, there is the concern that research will increasingly focus on for-profit areas, causing certain areas that do not make as much money to be neglected, even though they are in the public interest. In examining enabling legislation such as the Hatch Act[19] and the Morrill Act[20] from the 1860s and 1870s, it becomes apparent that land-grant universities were established to counter what were called the "elite institutions," that is, the Harvards, Yales, and other Ivy League schools. The mission of the land-grant universities is to provide public education for the broad masses of people in a democratic way and accomplish research in the public interest.

One question that naturally follows, then, is, what potential problems come with increased private funding? Some of the increase in private funds is due to declining federal funds. Moreover, what are some of the problems with private-sector funding of public university research projects? One problem is the emphasis on profit making and the ensuing neglect of research in those "nonprofitable" disciplines, with universities behaving more like for-profit companies. In addition, the deals between companies and universities involving patented research could restrict the flow of scientific information and research tools among scientists.[21] Scientists will delay publication and will decline to share data with others. And although there have been very few studies on this, some studies are emerging.

In a 1997 study, the *Journal of the American Medical Association* (*JAMA*) surveyed 2,167 university scientists.[22] Almost one-fifth, 19.8 percent, delayed publication of their scientific results for over six months to protect proprietary information. Another 8.9 percent admitted to refusing to share results with colleagues. These figures are representative of responses that were actually being reported. In my personal experience, positive data are published very quickly in prominent journals, while negative results remain unpublished or are published much later.

One example of an "early success story" is recombinant bovine growth hormone (rbGH) in the mid- to late 1980s. By 1985, researchers were reporting that injecting cows with rbGH increased milk yields by up to 40 percent.[23] At the time, unpublished studies were reporting that there was no evidence that bovine growth hormone had any adverse effect on animals or any human health implications. The Food and Drug Administration (FDA) had concluded in 1985 that milk from treated animals was safe to consume and had allowed milk from experimentally treated animals to go into the food chain. However, information from the University of Vermont leaked that there had been serious problems with mastitis[24] in one rbGH field study in Vermont. The field study had ended in 1988. In the next couple of years, the only animal health data reported at scientific meetings from this trial focused on somatic cell levels found in the milk; somatic cell counts of more than one million per milliliter of milk indicate clinical mastitis. While published data on these somatic cell counts suggested there was no problem at all, inside whistleblowers claimed that clinical mastitis had been a problem.

I went to Vermont in 1991 to testify to their House and Senate Committees on Agriculture about potential human health concerns associated with the use of rbGH and to urge them to request the animal safety data from the university. The Vermont House and Senate Committees on Agriculture had heard of the rumors and asked the university to turn over animal health data from the university's rbGH trials. In October 1991, the university, under orders from Monsanto, refused to release these data. The university argued that they were protecting the First Amendment rights of scientists to be able to publish information. The argument had little validity, as the field study had ended in 1988, and by late 1991, the study still had yet to be published. The University of Vermont is a land-grant university, and there is a sense that the motivation of the land-grant university was always to have the public interest as its highest goal, which is why this situation was so unsettling for the Vermont legislature.

But the issue did not go away. The Government Operations Committee, with jurisdiction over the FDA, sent a letter to the FDA requesting information about the animal health data from the Vermont trials. The FDA responded to the committee with a letter, dated December 12, 1991, that revealed that, in a Vermont trial, over 40 percent of the rbGH-treated cows had to be treated for mastitis, compared to less than 10 percent of the control cows (9 of 21 rbGH-treated cows versus 2 of 21 control cows).[25] In 1992, Monsanto refused to release the data to the General Accounting Office (GAO),[26] the watchdog arm of Congress.

The study finally published in the December 1992 issue of the *Journal of the Dairy Science*[27] found that, compared to control cows, rbGH-treated cattle were four times as likely to be treated with antibiotics for mastitis (8 vs. 2); had more than seven times as many cases of mastitis (29 vs. 4), with the average length of treatment for a case of mastitis being almost six times longer (8.9 days vs. 1.5 days), and had more than seven times as much milk discarded due to mastitis (73 kg vs. 10 kg).

Perhaps the only reason the information was ever published was because of the controversy surrounding this case. While the study was published in December 1992, it is difficult to believe that data collected in 1987–1988 required that much time to analyze. This lengthy time gap hints at the importance of secrecy in this issue and the desire to downplay any negative results as long as possible.

There is also the argument that money going into research can bias the findings. Scientists repeatedly say, "Just because I get money from a pharmaceutical company or from a chemical company will not skew any of my research findings." In fact, scientists often get angry if that question is raised, because they feel like their work ethic is under attack. I have experienced professors getting visibly upset with me for even inquiring about their funding sources. But medical studies done in the past couple of years show that funding source does have an impact.

In 1998, Deborah Barnes and Lisa Bero investigated research on secondhand, or passive, smoke.[28] They looked at 106 review articles and were concerned about the fact that the EPA, the National Institutes of Health, and the International Agency for Research in Cancer had all agreed that secondhand smoke caused negative health impacts including cancer, yet some review articles suggested that secondhand smoke wasn't harmful. Barnes and Bero wanted to understand the reasons for such findings. They hypothesized that review articles concluding that secondhand smoke is not harmful would tend to be poor in quality, published in non-peer-reviewed symposium proceedings, and written by scientists with tobacco industry affiliation.

The researchers looked at several variables and had rigorous ways for determining whether authors were industry or tobacco affiliated or not. Using stringent criteria, they found that the only factor associated with concluding that secondhand smoke is not harmful was whether the author was affiliated with the tobacco industry. Of the 106 review articles analyzed, 37 percent found that secondhand smoke was not harmful, and of those reviews, 74 percent were from industry-affiliated scientists. The number of tobacco-affiliated lead authors that found passive smoking harmful was only 2 out of 31, or 6 percent; 29 out of 31, or 94 percent found passive smoking not harmful. If you look at the 75 authors/articles that were not tobacco affiliated, 65 people, or 87 percent, found that passive smoking was harmful, and 10 people, or 13 percent, found that it was not harmful. Statistically, there is a very significant difference between the two. The correlation suggests that if you get money from tobacco companies, you are far more likely to find that passive smoking is not a problem.

In another study from 1996, Mildred Cho and Lisa Bero looked at 127 articles from symposia and 45 peer-reviewed journals. They found that 98 percent, or 39 out of 40, of the articles that had drug company support favored the drug, while only 79 percent, or 89 out of 112, articles without drug company support favored the drug. They concluded that "articles with drug company support are more likely than articles without drug company support to have outcomes favoring the drug of interest."[29] In other words, the funding source does influence the findings.

In 1998, another study was conducted regarding the calcium channel blocking controversy.[30] Initial studies claimed that calcium channel blockers were safe to use in treating cardiovascular disease. However, a number of studies published in 1995 raised concerns. One case control study suggested a possible association between the use of these calcium channel blockers to treat hypertension and increased risk of heart attack.[31] Another case control study of antihypertension medications in the elderly also found increased risk associated with calcium channel blockers.[32] These studies and others led to an intense debate in both the medical literature and the lay press.

A team of scientists from the University of Toronto decided to examine all the medical literature between March 1995 and September 1996 for articles written in English on the calcium channel blocker controversy.[33] They found a total of 70 studies, with the authors falling into three different categories. The authors were either for, neutral, or against calcium channel blockers. The supporters said that the good associated with channel blockers outweighs the bad. The neutral authors said, "On the one hand it does appear they are good for high blood pressure. On the other hand, there might be these problems." The authors that did not support calcium channel blockers primarily spoke of the problems. In examining the studies, the Toronto researchers found that supporters were significantly more likely to get money from the drug companies than the neutrals or nonsupporters. Of the authors that received money from a company that was manufacturing one of the calcium channel antagonists, 96 percent of the supporters, 60 percent of those who were neutral, and 37 percent of the authors that did not support the calcium channel blockers received funding. Again, these differences are statistically significant. If you look further and broaden the inquiry to include whether the researchers got any money from the pharmaceutical industry, 100 percent of the supporters, 67 percent of the neutral, and 43 percent of the nonsupporters received money from the pharmaceutical industry. The study concludes that "our results demonstrate a strong association between authors' published positions on the safety of calcium channel antagonists and their financial relationships with pharmaceutical manufacturers."[34] So, again, this larger-scale analysis suggests that the source of money does indeed have an impact on the outcomes of studies.

I find this conflict of interest fascinating and sad in many ways. These companies argue that even though the universities are receiving money from private industry sources, this receipt should not be a problem. Sheldon Krimsky at Tufts University is probably the national expert on this topic.[35] His 1998 study found a widespread conflict of interest with no public disclosure.[36] The study consisted of 789 articles that were all published in 1992 and written by authors from Massachusetts. The study found that about one-third of the lead authors in these articles had financial conflicts. Unfortunately, out of those 789 articles, none of the articles disclosed the conflict. Krimsky and his colleagues did another study in 1999 where they looked at 62,000 articles for corporate financial ties in very rough ways, and they found that disclosure was less than half a percent.

A national survey looking at disclosure of conflict-of-interest policies in biomedical research was published in November 2000.[37] The researchers surveyed the 127 medical schools and 170 research institutions receiving $5

million in total grants annually from the NIH and the NSF, 48 basic science
and clinical medicine journals, and 17 agencies, looking at their conflict-of-
interest policies. What they actually found was that the policies were all over
the board. Six percent of the institutions reported they had no policy. Among
the institutions that had policies, there were marked variations in the definition of management and conflicts, with 91 percent of the policies adhering to
the federal threshold for disclosure,[38] 9 percent with policies exceeding federal
guidelines, and only 8 percent with policies requiring disclosure to funding
agencies. Of these, only 7 percent had such policies requiring disclosure to
journals, and only 1 percent had policies requiring disclosure of information
to the relevant institutional review boards or research subjects. The people
performing the study were shocked.[39]

On January 30, 2001, the *Wall Street Journal* carried a front-page article entitled "Laboratory Hybrids: How Adroit Scientists Aid Biotech Companies with
Taxpayer Money."[40] What this entails is setting up a nonprofit that is connected to a for-profit, a scheme that might be particularly attractive to university researchers. The nonprofit then can get money from the NIH. The first
example discussed in the article concerned David Goldenberg, who had set
up a nonprofit called the Garden State Cancer Center, a center testing antibodies designed to detect or fight diseases. He also started a company called
Immunomedics Inc., a publicly traded biotechnology company which makes
antibodies that are then clinically tested at Garden State, and which may subsequently be commercialized. Dr. Goldenberg, the founder, chairman, and
CEO of Immunomedics, was given the first pick at the research from the
Garden State Cancer Center; furthermore, he had a 13 percent stake in
Immunomedics, worth over $100 million before the tech market decline. His
family and companies held another 10 percent. Goldenberg and his family at
the time were worth a couple hundred million dollars. Is this a nonprofit?
Eighty-six percent of Garden State's funds for the last four years, amounting
to approximately $30 million, have come from the NIH.

This type of hybrid nonprofit, one receiving federal funds and being
closely associated with a for-profit company, can be traced back to the Bayh-
Dole Act. With the dramatic increases in NIH funding in the past five years,
entrepreneurial scientists have found that the way to get government funding
is via the use of such hybrid nonprofits. NIH guidelines say that such nonprofit recipients should carefully "maintain their independence to pursue
their own missions without undue influence or restraint." This seems difficult when the head of the nonprofit is also head of a for-profit company. But
the heads of the for-profit companies argue that there's real independence:
"There isn't any real connection here. I'll take off my hat as the head of this
company when I'm negotiating with the other side of myself." The most troublesome aspect of the *Wall Street Journal* article was that the NIH was aware of
this "double-dipping" and responded by saying there was no legal reason to

ban such applications, as they don't enforce their guidelines and instead rely on the nonprofits to police themselves. That's like asking the fox to guard the henhouse. It goes without saying that the general public would be upset if they knew how much money companies receive from the NIH when they are already worth so much. Some say that this practice is part of the grantsmanship of the academic game, and so people look the other way.

These examples show that this influx of industry money into the land-grant universities is having an effect and that these are issues that need to be grappled with. And, as with all arguments, there are two sides. On the one hand, there is a need for more public funding allocated for agriculture research. Consumers Union has been lobbying every year on the Hill for an increase in public funding for agriculture research, but we have only been successful in obtaining small sums that go into integrated pest management or sustainable agriculture programs. We are always trying to increase funding for public universities because public universities can ensure the public interest and are providers of knowledge and services that the market cannot provide. However, it seems like there are some university administrators that see the university as a means for making more money, looking at it as a business rather than a university. These people are forgetting that the basic mission of universities is to explore questions whose answers are unknown, with the understanding that there is a freedom to publish unpopular results with no fear of reprisal from peers.

However, the problem is not only with the universities. The NIH is providing less and less money for true basic research and is encouraging universities to compete with industry. Setting up these fake nonprofits and for-profits seems to be a part of this phenomenon. Much has to change, and at least some of the debate will need to emerge out of the public sector.

It is not that industry should not fund universities or that private or collaborative work is never in the public interest. The problem is that industry support has gone to an extreme, and at the very least there needs to be absolute disclosure. This disclosure should occur not only within universities, but if anyone has funding from a corporate source, it should be made public knowledge for the sake of accountability. Although corporate funding does not necessarily mean that the research is biased, it is something that should be taken into account. Sources of funding are important, and if universities want to be seen as a neutral third party, they will increasingly be seen as biased without full and complete disclosure.

There was a time when a researcher would be ostracized if he or she tried to patent or otherwise make money from his or her research. This was because the whole ethic was that researchers were doing research for "knowledge's sake," not for the money. With the rapid commercialization of the university, this mindset is rapidly disappearing. Intuitively, one would think that the best science would come from people that are doing things because they are obsessed with and interested in it, not because they are chasing money or fame. When asked whether he would patent the polio vaccine, Jonas Salk responded, "What are you going to do next? Patent the sun?"

Following are Dr. Hansen's answers to questions posed during a discussion of his paper at the University of Illinois Center for Advanced Study.

Q: You draw a line between public and private research throughout your presentation. Why can't research that is profitable also be in the public interest, that is, for the general public good? Do we immediately assume that those two things cannot go together?

A: *No, not at all. I think that it is good to try to take work that is done at basic science institutions and get that out to the general public in the form of technology. What is disturbing with these university-industry connections is that we are starting to see scientists coming into various conflicts of interest. They have an economic stake in work they are doing, and if their research turns out positively, they might make themselves millions of dollars. They are no longer the disinterested academics.*

As I mention in the talk, you can look at studies that suggest that drug researchers who have an economic interest in or work for a company produce articles that tend to be much more positive about drugs produced by that company than researchers who are not getting money. The data do in fact show that money can influence.

Q: There is precedent in the hard sciences for researchers who work at universities to benefit economically from their work. And I am not quite sure why it should be a problem for people in the life sciences to do that. As they sit in their offices, they look across the campus at their colleagues who work in areas like electrical engineering and computers, and they see them making a profit from spin-offs from their research. Why should we say that people in life sciences, or anybody else, should not be able to do that?

A: *The problem may be the same; I am not sure. My background is in the life sciences, so it is easier for me to see when profit-driven research displaces other research in that area. Researchers decide to pursue industry-sponsored high-tech genetic engineering approaches, while more sustainable, ecologically based approaches are being ignored.*

Q: It costs a lot of money to do scientific research, and the government is obviously not providing adequate funds. Researchers have to go somewhere to get the money, so they are going to the private sector. If the government is not going to provide the money, then what choice do researchers have?

A: *I think that the government does need to provide more money for research that is in the public interest, but I also think it is up to the universities to change their policies. It may be okay for individual scientists to look and to try to get some funding from corporate or other sources, but the fact is that it is happening on such a large scale that universities themselves have been drawn into the competition. Universities themselves are deciding on a campuswide basis that they want to increase profit-making departments and reduce not-as-profitable departments. Consequently, some of these universities are deciding to shut down some of the humanities and other departments because they are not bringing in money. When administrators are concerned only with figuring out ways they can make money with their patents and other research, then the whole mission of the university starts to change. It would be one thing if there were a few scientists going outside to try to get this money and the people that run the university still kept the public interest at heart and were very stringent about disclosure. But universities are not doing that. They seem to be running to embrace higher-tech research as a way of generating funds, and the fact that these things are happening at the highest levels of the university is what makes it seriously problematic.*

NOTES

1. Federal Technology Transfer Act of 1986, Pub. L. No. 99–502, 2, 100 Stat. 1785–87 (codified as amended at 15 U.S.C. 3710a[a][1], [b][2]-[3] [2000]).

2. Eyal Press and Jennifer Washburn, "The Kept University," *Atlantic Monthly*, 285, no. 3 (2000), 41.

3. Press and Washburn, "The Kept University," 41.

4. Press and Washburn, "The Kept University," 41.

5. Press and Washburn, "The Kept University," 41.

6. Michael Hansen et al., "Plant Breeding and Biotechnology," *BioScience* 36 (1986): 29.

7. Paul De Bach, *Biological Control by Natural Enemies* (Cambridge, UK: Cambridge University Press, 1979).

8. De Bach, *Biological Control by Natural Enemies.*

9. See www.epa.gov/swercepp/ehs/profile/12002038.txt. Paris Green is an extremely poisonous, green powder sometimes used as a pesticide.

10. David J. Greathead and Jeff K. Waage, "Opportunities for biological control of agricultural pests in developing countries," World Bank technical paper no. 11 (Washington, DC: The World Bank, 1983).

11. David Pimentel et al., "Benefits and Costs of Pesticide Use in U.S. Food Production," *Bioscience* 28, no. 12 (1978): 772.

12. J. E. Lange and J. Hamai, "Biological Control of Insect Pests and Weeds by Imported Parasites, Predators, and Pathogens," chap. 28 in *Theory and Practice of Biological Control,* ed. Carl B. Huffaker and Paul S. Messenger (New York: Academic Press, 1976).

13. Sandburg, B., "The Ivory Tower IP Fix," *IP Magazine*, May 1999.

14. Press and Washburn, "The Kept University," 40.

15. Michael Hansen, *Escape from the Pesticide Treadmill* (Penang, Malaysia: Published by International Organization of Consumers Unions and Pesticide Action Network, 1988). See chap. 4.

16. "IITA's Research Wins the 1995 World Food Prize," www.worldbank.org/html/cgiar/newsletter/Mar96/4wfp.htm.

17. Stanley Ziemba, "FTC Approves Drug Giants' Merger," *Chicago Tribune*, December 18, 1996. The Novartis Pharmaceuticals Corporation was created in 1996 with the merger of Sandoz and Ciba.

18. See Cary H. Sherman and Steven R. Englund, "When the Feds Share the Tab," *Legal Times*, May 15, 1995, 21. An even more controversial arrangement had been established—and challenged—earlier between Scripps Research Institute and Sandoz, in which Sandoz was first awarded the right to license all of Scripps' inventions over the course of ten years.

19. See 7 U.S.C. 361a et seq., 2000. The Hatch Act of 1887 authorized federal-grant funds for direct payment to each state that would establish an agricultural experiment station in connection with the land-grant college established under the provisions of the Morrill Act of July 2, 1862, and of all supplementary acts.

20. See 7 U.S.C. 301 et seq. The Morrill Act of 1862 established the land-grant university system. The legislation introduced by U.S. Representative Justin Smith Morrill (VT) granted to each state 30,000 acres of public land for each senator and representative under apportionment based on the 1860 census. Proceeds from the sale of these lands were to be invested in a perpetual endowment fund that would provide support for colleges of agriculture and mechanical arts in each of the states.

21. Paulette Campbell, "Pacts between Universities and Companies Worry Federal Officials," *Chronicle of Higher Education* 36 (1998): A37.

22. David Blumenthal et al., "Withholding Research Results in Academic Life Science: Evidence from a National Survey of Faculty," *Journal of American Medical Association* 227, no. 15 (1997), 1224.

23. Robert Kalter et al., "Biotechnology and the Dairy Industry: Production Costs and Commercial Potential of the Bovine Growth Hormone," AEA Research 85–20, Cornell University Department of Agriculture Economics, 1985.

24. Michael Hansen, "Potential Animal and Human Health Effects of rbGH Use," testimony before the FDA's Veterinary Medicine Advisory Committee, March 31, 1993.

25. K. Holcombe, letter from the acting associate commissioner for legislative affairs, FDA, to the Hon. Bernard Sanders, dated December 12, 1991.

26. In 2004, the GAO changed its name to Government Accountability Office.

27. Alice Pell et al., "Effects of a Prolonged-Release Formulation of Sometribove (n-Methionyl Bovine Somatotropin) on Jersey Cows," *Journal of Dairy Science* 75 (1992): 3416.

28. Deborah Barnes and Lisa Bero, "Why Review Articles on the Health Effects of Passive Smoking Reach Different Conclusions," *Journal of the American Medical Association* 279, no. 17 (1998): 1566.

29. Mildred K. Cho and Lisa Bero, "The Quality of Drug Studies Published in Symposium Proceedings," *Annals of Internal Medicine* 124 (1996): 485.

30. Henry Stelfox et al., "Conflict of Interest in the Debate over Calcium-Channel Antagonists," *New England Journal of Medicine* 336, no. 2 (1998): 101.

31. B. Psaty et al., "The Risk of Myocardial Infarction Associated with Antihypertensive Drug Therapies," *Journal of the American Medical Association* 274 (1995): 620.

32. M. Pahor et al., "Long Term Survival and Use of Antihypertensive Medications in Older Persons," *Journal of American Geriatric Society* 43 (1995): 1191.

33. Henry Stelfox et al. "Conflict of Interest in the Debate over Calcium-Channel Antagonists."

34. Henry Stelfox et al. "Conflict of Interest in the Debate over Calcium-Channel Antagonists."

35. Sheldon Krimsky is the professor of urban and environmental policy at Tufts University. Professor Krimsky's research has focused on the linkages between science/technology, ethics/values, and public policy.

36. Sheldon Krimsky et al. (1998). "Scientific Journals and Their Authors' Financial Interests: A Pilot Study," *Psychotherapy and Psychosomatics* 67, no. 4–5 (1998): 194.

37. S. Van McCrary et al. "A National Survey of Policies on Disclosure of Conflicts of Interest in Biomedical Research," *New England Journal of Medicine* 1612 (2000): 343.

38. McCrary et al., "A National Survey of Policies on Disclosure of Conflicts of Interest in Biomedical Research." The federal guidelines require that if a person has $10,000 in annual income or equity in a relevant company or 5 percent ownership in a pharmaceutical company, then this fact must be disclosed.

39. McCrary et al., "A National Survey of Policies on Disclosure of Conflicts of Interest in Biomedical Research." "We are disturbed that only 7% of the institutions and only 43% of the medical and scientific journals that we surveyed required disclosure of financial interest in published reports on research. These findings suggest that readers who assume that financial conflicts would be uniformly disclosed are mistaken."

40. Chris Adams, "Laboratory Hybrids: How Adroit Scientists Aid Biotech Companies with Taxpayer Money," *Wall Street Journal*, January 30, 2001, A1.

12

The Governmentalization and Corporatization of Research

Toby Miller

In this chapter, I will address the governmentalization and corporatization of research universities[1] in the United States and its links to labor and the new humanities, aka cultural studies. I contend that while universities have never been free from direct and indirect influences of government and business, the contemporary era is witness to a decrease in the relative autonomy granted to U.S. research scholars since the 1960s. Cultural studies as a field is in fact a response to these tendencies, and also a means of managing "left-ism."

Medieval European universities, such as the University of Paris, were no safe harbors of disinterested inquiry; that is, the original intent of the university was not to pursue "knowledge for knowledge's sake." Rather, they were born as sites for ruling and emergent classes to exercise economic and political power, where their successors could learn to utilize knowledge as an addendum to scaffold the cultural climate of the time. Universities grew in power as a struggle between clerical and secular knowledge and power developed. This struggle was frequently taken to the streets in the form of violent protests against city rulers. When national governors intervened, it was mostly on the side of the emergent class against the city. In the eighteenth and nineteenth centuries, the target group of the university changed, as a system of education for all young people developed across Europe as a means of encouraging its emergent industrial urban proletariat to be obedient, willing participants in society. This system inevitably led to broader access to universities as well (Boren 2001, 9, 14, 22, 27). This developing system represents the emergence and development of educational governmentality and is also linked to corporatization, terms which I will hereafter define.

GOVERNMENTALITY

Michael Foucault: On the Historicity of Governmentality

Michel Foucault offers a history of modern sovereignty via the "barbarous but unavoidable neologism: governmentality." This term, "governmentality," coined by Roland Barthes, was used to describe contemporary responses of the state toward economic change—a form of contingent responsibility (1973, 130). Foucault then historicizes the term to identify a series of problems addressed at different moments in European economic and political organization. He begins with five questions posed throughout the sixteenth century: "How to govern oneself, how to be governed, how to govern others, by whom the people will accept being governed, and how to become the best possible governor." These questions emerged from the displacement of feudalism by the sovereign state, the Reformation, and the Counter-Reformation. Daily economic and spiritual government came up for redefinition as the state sought to normalize itself and society. Religious authority, embroiled in ecclesiastical conflicts, lost its legitimacy to vouchsafe the sovereign's divine right of rule. The monarch was then gradually transformed into more of a manager rather than the embodiment of eminence (Foucault 1991, 87–90).

From that time on, governing required a double movement eventually producing a comprehensive system of education. Initially, sovereign powers discovered how to run their own lives and treated their dominions according to these lessons. In similar fashion, fathers ran families like principalities and trained their "minions" to carry domestic docility and industry into the social sphere. This backward and forward motion between the public and private, of imposed and internalized norms shuttling between work and home in search of civic peace and control, formed the basis of public schooling. Pedagogy extrapolated from the self-knowledge of rulers to the rule of others.

We might see this as the economization of government, a complex movement between self and society in search of efficiency and authority. With such upheavals as the Thirty Years' War and rural and urban revolts during the seventeenth century, new modes of social organization inevitably arose. In eighteenth-century Europe, the concept of "the economy" spread beyond the domestic sphere. What had been a managerial invention, dedicated to developing "correct" conduct, had transformed itself into a description of the social field. By now, the government of territory was secondary to the government of things and the social relations between them. Government was conceived and actualized in terms of climate, disease, industry, finance, custom, and disaster—literally, a concern with life and death, and what could be calculated and managed in between them. Wealth and health became social goals, to be attained through the disposition of capacities across the population—"biological existence was reflected in political existence," through the work of "bio-power." Bio-power "brought life and its mechanisms into the realm of explicit calculations and made knowledge-power an agent of transformation of human life." Bodies were identified with politics, because managing them was part of running the country. And this history is still relevant to contemporary life. For Foucault, "a society's 'threshold of modernity' has been reached when the life of the species is wagered on its own political strategies" (1991, 97, 92–95; 1984, 143).

The foundations of classical political economy date from around the seventeenth century and are generally associated with a libertarian championing of the market. But as Michael J. Shapiro's (1993) study of Adam Smith has shown, the very founder of the discourse theorized sovereignty beyond exhibiting and maintaining loyalty. Government was required to manage "flows of exchange within the social domain" (11). Both Smith and the physiocrats identified the transformation in the status of government from legitimacy to technique, specifically in the ability to distinguish "what is free, what has to be free, and what has to be regulated," notably in the areas of crime and health (Foucault 1994, 124–25). Science and government combined in new environmental-legal relations under the signs of civic management and economic productivity. In 1855, Achille Guillard merged "political arithmetic" and "political and natural observations" to invent "demography," which had been on the rise since the first population inquiries back in seventeenth-century Britain. The new knowledge codified five projects: reproduction, aging, migration, public health, and ecology (Fogel 1993, 312–13).

The critical shift here was away from an autotelic accumulation of power by a sovereign monarchy and toward its decentralization of power to the general populace via the formation of self-sustaining, marketable skills. This resulted in a necessity to equip the citizenry with the capacity to produce and consume things, and insistence on freedom in some compartments of life and subservience in others (Foucault 1994, 125). At the same time, philanthropic endeavors were also developing in Europe—the beginnings of today's third sector, in between the private and the public. Neither profit based nor state based, it was occupied by a small circle of people interested in social reform and operating beyond the pressing, self-interested norms of politics, yet in a governmental mode (Donzelot 1979, 36, 55–57, 65).

This new duality—empire and economy—expanded the purview of government beyond the sovereign monarchy and households. Populations displaced princes as sites for accumulating power, just as national economies displaced homes as sites of social intervention and achievement (Foucault 1991, 98–99). Educational policies became part of the set of expected domestic responsibilities. For instance, at the turn of the twentieth century, the British introduced an "education for all" policy that was founded in a (frequently problematic) knowledge of difference as much as in utility. The subsequent

Education Act of 1902 mandated school-pupil visits to museums to encourage encounters with other worlds (Coombes 1994, 124).

In sum, the rise of modern capitalism is connected to the rise of the sovereign state, which was concerned about delivering a subservient but healthy labor force to the business sector as well as others. At this point, then, the emergence of the health and physical well-being of the population in general became one of the essential objectives of political power. The entire "social body" was assayed and treated for its failings. Governing people primarily and most importantly meant obeying the "imperative of health: at once the duty of each and the objective of all" (Foucault 1978, 277). So while education was engineered to enable government and business, it was also to endow the public with a sense of self and history that welcomed public critique.

Higher Education in the United States

The university in the United States has taken some of these elements and also incorporated a few key things that are very characteristically American. The United States has constructed a system of university funding, policy, and management that in fact blends governmentalization and commodification. Since the 1830s, when the first waves of white interclass European immigration began, higher education has performed dual functions: integrating the U.S. population (Aronowitz 2000, 5) through national "myth making" and generating new practices and knowledge for use by states and businesses. Of course, this was more a tale of uneven development rather than an even turn toward governmentalization and commodification. By the 1850s, however, with the rapid industrialization in the United States, the new chiefs of industry envisioned partnerships with higher education as a means of developing a skilled workforce.

The Republican Party, under President Abraham Lincoln, enabled this alliance via the land-grant system. Technocratic from its inception, the land-grant university flowered as a business mimic by the turn of the century, as corporations were placing more and more faith in applying science to their endeavors in electromagnetism, geology, chemistry, and electricity for communications technologies. By the 1920s, Harvard had its business school, NYU its Macy's-endorsed retail school, and Cornell its hotel school (Pietrykowski 2001). It is no wonder that Thorstein Veblen referred to U.S. universities as "competitors for traffic in merchantable instruction" (quoted in Pietrykowski 2001, 299). His old-fashioned words remain very accurate in their diagnosis. The two world wars provided additional pump priming and premia on practicality from the federal government; in fact, the big research schools even expanded their capacity during the Depression (Aronowitz 2000, 16, 18–20). It was during this time when higher education finally started becoming more accessible to the public (Aronowitz 2001, 94).

In the research domain, the notion of mutual investment endorses partnerships between state, college, and industry. Such relationships merit scrutiny rather than an amiable blind faith. The cold war "became the instrumentality of a vital national economic policy," acting as a catalyst for expanding policy and institutional practices, as evidenced in the growth of the university via increasing federal and state subsidies (Lewontin 1997, 7). Since then, considerable effort has gone into clarifying the significance of tailoring research priorities to contemporary political parties and corporations, aka "pork barreling." Major research institutions like Harvard and Stanford literally have had dozens of formal corporate partners since the 1980s (Aronowitz 2000, 44). Ralph Nader's Center for Universities in the Public Interest was set up because of such concerns, which are evident to Derek Bok, Robert Nisbet, and others—former supporters of government-college-industry relationships who have experienced the obstacles that such relationships can pose to honest research outcomes. The complications are obvious in currently hot topics such as bioethics, but there are issues for other fields, like anthropology, which is implicated in the unfolding controversy over the Yanomami in Venezuela and Brazil, involving the well-known U.S. anthropologist Napoleon Chagnon. Another example is psychology departments, which frequently require undergraduates to present themselves as research subjects as a condition for enrollment, with the results, publication, presentation, or commodification of no tangible benefit to them.

Much academic participation in democratic government also proves to be problematic. Consider language-spread policies and the part that linguists play in this, the work of economic advisers,[2] political scientists,[3] biomedical researchers in their relations with pharmaceutical companies, public-relations consultants, sociobiologists and their defenses of male sexual violence, and so on. The very existence of communication research raises questions of ideological distortion, given the formation of the discipline under the sign of war, clandestine government activity, and later corporate and foundation support (Simpson 1996). The same could be said of applied areas in political science and public administration. These policy-oriented sciences, originally conceived as bridges connecting democratic and executive action, have been reduced to an "unrepresentative expertise" lacking articulation with the everyday. Thomas Streeter (1996) points out that in the United States, "policy" frequently connotes a procorporate position that turns highly contestable positions into absolutes, with consultant professors simultaneously performing objectivity and applicability (14, 133, 136).[4] With the exception of cultural anthropology, the social sciences have become so policy oriented in their ends, and science oriented in their method, that they "have largely lost their critical character." Some say that they have even given up on the task of socialization, other than instruction in possessive individualism (Aronowitz 2000, 4, 40).

Of course universities have key links to think tanks. Consider those who promote the interests of capital and the status quo from the well-heeled offices of RAND, Olin, the American Enterprise Institute, or Brookings. For example, RAND's contributions to Vietnam were crucial to the displacement of behavioralism by rational choice as a favored method in the social sciences, law, and policy. All of this was achieved by interfering with the lives of Southeast Asian citizens during the Yanqui imperialist war. Initially, the alleged insights of U.S. behavioralists recommended development, village infrastructure, and military services to win over the Vietnamese. When this failed, rational choice arrived, claiming that maximizing the pain of the opposition would make them cower. This meant that the certainty of pain and horror resulting from using napalm would lead to victory of freedom over totalitarianism.

This moribund history long predates today's concerns about financing the U.S. research university since we lost the relatively disinterested cold war stimuli to big science. But it appears as though governmentalization and commodification have recently merged in their concerns and methods. Today, Congress provides more than a billion dollars in direct grants to universities, in addition to peer-reviewed funds available through the NSF and the NIH. This is in contrast to corporation funding, which was recorded at $850 million in 1985, increasing to $4.25 billion in 1995 (Poovey 2001). The multinational pharmaceutical corporation Novartis currently funds more than a third of the activities of the Plant Biology Department at the University of California at Berkeley. MIT's media laboratory is seen by many outsiders as a playpen provided by corporations for well-meaning but apolitical graduate students working with implicit and explicit theories of possessive individualism. And industrial research parks now dominate work at research schools like the universities of Texas, Massachusetts, and North Carolina, as well as Duke and Stanford. However, not all such ventures are supported by private money. The NSF established dozens of engineering research centers in the 1980s with the expectation that they would form partnerships between corporations and higher education, which have effectively functioned more like ongoing public welfare for "entrepreneurs" (Rhoades and Slaughter 1998, 36).

U.S. universities are increasingly businesslike institutions thanks to the Bayh-Dole Act of 1980, which permitted educational nonprofit institutions to own and commercialize inventions. At times, universities have even taken legal action against their own researchers to maximize their profit. Prior to 1980, research universities collectively accounted for about 250 patents a year. Now the figure is closer to 5,000, along with approximately 3,000 new companies that have emerged as a consequence of the legislation (Blumensty 2002; Poovey 2001). In 1999, the top hundred research schools received $641 million in royalties, a $500 million increase in just four years (Goldschmidt and Finkelstein 2001). The idea of working for the public interest has been forgotten, erased by amendments to state laws that have quietly exempted publicly funded scientists from the same "conflict-of-interest" responsibilities as refuse workers or personnel officers (Rhoades and Slaughter 1998, 39).

While it is difficult to ascertain how much of the federal research money allocated to universities also funds corporate-oriented research too, Washington continues to provide mammoth direct and indirect subsidies. Looking at table 12.1, we can see great consistencies through the 1990s in the proportion of money spent for direct and indirect corporate research welfare. Of course almost all of it goes to the sciences, with small amounts allotted for the social sciences, and essentially nothing for the humanities.

Diverting our attention away from the research sector, the liberal arts model of education takes on the difficult task of trying to equip students with a democratic inclination that respects knowledge *of* a topic and *for* a purpose, rather than simply knowledge from a particular person. The model gives us a system based on a discourse of professionalism rather than charisma. Of course, this liberal arts model also incorporates the concept of human capital, in that there should be a mutual investment of time, money, and training by both society and subject to create able-minded technical employees and willing patriots. To this end, we have also seen the formation of the idea of higher education as an industry and students as consumers. Thus, Bruce Johnstone, a former chancellor of the State University of New York, offered the concept of "learning productivity" as part of the process of students beginning to "assume greater personal responsibility for their learning" (quoted in Martin 1998, 9).

We can see a tendency across the entire degree-granting sector to transfer the cost of running schools away from the government and toward students. In 1980–1981, the three levels of government accounted for 48.3 percent of the funding, whereas in 1995–1996, it was 38 percent. This trend toward relying on tuition for university funds has doubled student debt from 1992 to 2000 and has reinforced the idea of the student as a consumer rather than a citizen (Chaker 2002).

As part of this logic, universities engage in student evaluations of faculty. Today, such methods are used by 95 percent of all departments (Rhoades and Sporn 2002, 360). Of course, underpinning this logic, the idea of a market in which the employer (either the university or the university administrators) allocates resources (teaching) according to what consumers (students) want is utterly fantastical. In a country where at any given time there are 50,000 undergraduates majoring in English and 4,000 in physics (Martin 1998, 13), it would seem there would be better labor conditions and benefits for English professors than physicists. Yet this is not the case, since demand is about corporatization by paymasters outside the academy, not necessarily involving the desires of students per se (see table 12.2).

Table 12.1. Federal funds for research, development, and related plant 1991–1999 in millions of current dollars (rounded to whole numbers)[1]

1991	1992	1993	1994	1995	1996	1997	1998–EST	1999–EST	CATEGORY
64,292	65,719	68,386	68,336	68,410	67,756	70,892	71,780	73,150.4	OUTLAYS
28,490	31,754	31,777	32,748	32,672	31,498	32,646	33,540	33,166	CORPORATE
13,772	14,126	14,823	15,121	15,507	15,391	16,260	16,844	18,065	EDUCATIONAL

[1] Includes federally-funded research centers. The disparity between outlays and the sum of moneys allocated to corporations and universities lies in sums provided to state and local governments plus not-for-profit, non-educational bodies.
(Source: Federal Department of Education *Digest of Education Statistics,* 1999)

Table 12.2. Current university fund revenue sources (1980–1996) by percentage

1980–81	1985–86	1989–90	1990–91	1991–92	1992–93	1993–94	1994–95	1995–96	SOURCE
21	23	24.3	25	25.7	26.5	27.1	27.2	27.9	Tuition
14.9	12.6	12.4	12.2	12.3	12.3	12.3	12.3	12.1	Federal government
30.7	29.8	27.5	26.4	25.1	24.1	23.4	23.4	23.1	State government
2.7	2.5	2.6	2.6	2.6	2.6	2.8	2.7	2.8	Local government
4.8	5.4	5.6	5.6	5.6	5.7	5.7	5.7	6.0	Gifts, grants, contracts
2.1	2.3	2.3	2.2	2.1	2.1	2.0	2.1	2.3	Endowments
23.9	24.4	25.4	26.1	26.5	26.6	26.6	26.5	25.7	Sales and others

(Source: Federal Department of Education *Digest of Education Statistics,* 1999)

This dependence on private sources is a kind of mimetic fallacy, a process where both governments and university administrators make corporate life into their "desired other." This not only makes for untimely influences on the direction of research and teaching, but on the very administration of the university, which is increasingly prone to utopian notions of "excellence" and "quality control." In short, academic institutions are coming to resemble the institutions they now serve, so much so that colleges have now been transformed into big businesses. Major research institutions, particularly private ones, are also landlords, tax havens, and R & D surrogates rolled into one, with the administrators and fundraisers lording it over the faculty. Deconal apparatchiks have replaced the era of faculty governance. University bureaucrats are transitioning toward a more CEO-like stature, exemplified in the vast redistribution of resources in reducing expenditures on instruction and increasing expenditures on administration (Rhoades and Slaughter 1998, 58).

Again, academic governmentalization has a long history. Regional accreditation offices vouching for the quality of U.S. degrees have been in place for over a century. Yet since the 1970s, we have seen an increase of performance-based evaluations conducted at the departmental and administrative level, rather than in terms of the standard overall quality of the university. Such systems directly link budgets to outcomes, in keeping with the prevailing beliefs of public-policy mandarins and their relentless quest to conduct themselves like corporate elves manqués. As successive superstitions came along—in the 1990s, it was "total quality management"—administrators fell in love with these beguiling fads. All the while, faculty-student ratios worsened while reporting, surveillance, and administration developed in size and power (Rhoades and Sporn 2002, 359–62, 366–67; Sora 2001). In reality, many of us who have actually worked for businesses and the government know what inefficient institutions they are. That brings us, of course, to the place where universities create value as a source of trained labor.

Academic Labor

I will look at this trend by examining a particular administration's managerial response to a struggle that took place from 1999 to 2002 concerning graduate-student employees' demands for industrial representation. Catharine Stimpson (2000), dean of New York University's Graduate School of Arts and Sciences, published an op-ed on this fiasco in the *Chronicle of Higher Education* entitled "A Dean's Skepticism about a Graduate Student Union." Her remarks are representative of widely held sentiments in the academy that are implacably opposed to student-employee unionization, and also the opportunity for students to express their democratic views on this matter.

Stimpson's article begins with the claim that unionization advocates an adversarial rather than a collegial means of governance. At NYU, which is fervently antiunion in general, such difficulties do in fact exist. But they arise from the administration's implacable opposition to industrial organization. She makes this claim without any outside, corroborating evidence. And while there may well be profound differences in terms of standpoint and interests generated by the different divisions of labor, these differences do not always color the interactions between employers and employees. As for the claim that the current modus operandi is collegial and not adversarial, I shall address that a little later.

Second, Stimpson asserts that while graduate students "do valuable work . . . they aren't employees." This reinforces the picture that graduate students are discounted labor, allowing these institutions to operate at a fraction of the cost that would be incurred for proper salaries being paid in return for the discharge of innumerable professional duties. The only relevant national data suggests that, for example in the humanities, graduate student labor sustains up to 42.5 percent of introductory classes and between 7 and 34 percent of all undergraduate instruction (Cox 2000, A12). If graduate students are not employees, perhaps the NYU administration might care to calculate and publish the cost of substituting their work for people paid at rates set by the market or collective bargaining, rather than for graduate students to be paid via "stipends."

Third, and perhaps most directly relevant to our concerns here, Stimpson claims that the industrial model generated by union membership cheapens the historic mission of the university. But this begs the question of where that process of industrialism began and how it is currently managed. The short answer lies in the process of governmentalization and corporatization that the university as a whole has been undergoing already. As I have shown, they have long been in train, albeit via unevenly developing processes and struggles. The presence of corporatization today is the outcome of the university's adherence to forms of funding and social influence connected with providing R & D services to government and business, borrowing fashionable forms of business management from corporations (i.e., practicing divide-and-conquer forms of administration to centralize power, already undermining the collegiality and autonomy that has traditionally characterized the university).

Fourth, Stimpson quotes an anonymous person complaining that graduate students are "damn well paid," showing a shocking disregard for the cost of living in New York City in terms of expenditures like health care, housing, and basic subsistence. Such shallow analyses reveal the indifferent bureaucratic way the administration views graduate students, simply as "single cogs in this big university machine."

Fifth, Stimpson maintains that looking at different situations with unions, such as the United Automobile, Aerospace and Agriculture Implement Workers of America (UAAAIWA), as representative of student concerns over these questions would stifle unbiased debate and influence decisions made concerning this issue at NYU. But that was not the intent of the union, the National Labor Relations Board (NLRB), or anyone else for that matter. Mandatory collective bargaining does not typically include such issues, because it requires the agreement of both sides to be included in negotiations. The idea that there would be a loss of "shared and collegial academic governance" presumes that such governance already exists at NYU without presenting any evidence pointing to that fact.

Sixth, Stimpson says that graduate students are a transient population and as such should not be permitted to vote on matters that have an effect on others. One might bear in mind that similar arguments were made against women workers gaining similar representation on the grounds that they, too, were a transient population.

Finally, on a similar note, Stimpson objects to the NLRB's exclusion of certain students from voting. It seems incredibly bad form to make this point, since this condition was passed because of NYU's own claims on this issue, which were then backed up by both the UAAAIWA and the NLRB—that students funded by professorial grants to undertake collaborative work that directly addressed their dissertation topics were not undertaking labor on behalf of the university, which would include things like photocopying course outlines or grading papers.

Early in 2000, the NLRB voted in favor of the unions and against NYU. An election was held, giving students the choice to decide whether they wished to be represented by the UAAAIWA or not. While the NYU administration did have some support among the faculty, a significant number opposed. One hundred and seventy faculty members signed a petition requesting that the university not appeal the NLRB's decision that graduate-student employees at NYU could vote determining whether a majority favored collective bargaining machinery as a means of financial amelioration. While many among the 170 faculty members supported the principle of unionization, there were those who were opposed. But all were horrified by the automatic denunciation of the right to vote—that the democratic privilege of self-expression concerning the desirability of union representation should be denied to a financially disadvantaged but intellectually, administratively, and professionally important fraction of the university population. But the election results were then sealed pending the outcome of an appeal by NYU's administration. NYU was, of course, aided in its struggle by the Ivy League schools, which expressed their anxiety that this petit bourgeois arriviste might lose the case, setting a precedent for all graduate students in graduate schools. NYU lost that appeal, and a contract was signed. When the Republican Party restructured the NLRB, its conservative majority reversed the decision.

The entire affair laid open the desire of elite schools to prevent graduate-student employees from expressing their views on a key topic—namely, the right to organize. Whatever one's views are on unionism, this was a shocking breach. NYU's antidemocratic conduct has seen the self-styled "global university" attract global and national condemnation for its authoritarianism from *The Economist* (Pupil Power 2000), the *New York Times* (Unions and Universities 2000), and *Doonesbury* cartoons, along with New York metropolitan and state legislators. As their notoriety spread, the administration and its antidemocratic confreres looked increasingly uglier and lonelier, finally losing the legal battle. However, it was an interesting coincidence that the only nontenured faculty member who testified on behalf of the graduate-student employees before the NLRB, Joel Westheimer, was soon coincidentally fired by NYU. The university was charged with an offense and subsequently settled out of court with Professor Westheimer in mid-2002.

RADICAL DEMOCRACY AND THE HUMANITIES

Of course, contrary to Stimpson's suspicions, connections between university's student unions can improve the quality of research and teaching—but in the context of applied radical democracy, not corporate administrative hegemony masked behind disinterested inquiry.

I am not fantasizing here about a world distant from applied intellectual labor. When we think about oppositional theory, some of the names and faces that come to mind are people such as the Italian semiotician and novelist Umberto Eco, linguistics professor and corporate-media critic Noam Chomsky, French philosopher Jean-François Lyotard, and the Argentinian-Mexican anthropologist Néstor García Canclini, to name a few. Some of their most famous work was born out of cultural consultancy and applied research: Eco's TV semiotics (1972) was a project undertaken in the 1960s for Italy's state broadcasting; Chomsky's transformational grammar (1965) was funded by the Joint Services Electronics Programs Department of the U.S. military; Lyotard's report on postmodernism (1988) was written for the government of Quebec; and García Canclini's theory of hybridity (1990) was derived from a report on indigenous crafts. These have all been crucial to the development of the new humanities—cultural studies—and a variant of that in cultural policy studies.

Yet the interdisciplinary nature of cultural studies and its focus on identity have been referred to by sources like the *Times Literary Supplement* as the "politico-intellectual junkyard of the Western world" (Minogue 1994, 27), while Chris Patten, a former Conservative Party member of the UK Parliament and the last governor of Hong Kong, calls the humanities a "Disneyland for the weaker minded" (quoted in Morley 1998, 478), and leading bourgeois economic maven Jagdish Bhagwati (2002, 3) is convinced that cultural studies is, in fact, largely to "blame" for global grassroots activism against globalization. On the other hand, Virginia Postrel, former editor of the right-wing libertarian *Reason* magazine, described cultural studies in a 1999 op-ed piece for the *Wall Street Journal* as "deeply threatening to traditional leftist views of commerce," because its notions of active consumption are similar to the sovereign consumer associated with the right, that "the cultural-studies mavens are betraying the leftist cause, lending support to the corporate enemy and even training graduate students who wind up doing market research."

On the left, cultural studies' concerns with identity and consumption have led to accusations of falling from the grace of "real" politics. I often think of the character in Don DeLillo's novel *White Noise*, whose complaint about his university is that "there are full professors in this place who read nothing but cereal boxes." *New Yorker* journalist Adam Gopnik has accused radicals in the United States of being overcommitted to abstract intellectualism and the assumption that "consciousness produces reality," such that the "energy on the American left is in cultural studies, not health care" (1994, 96). To this, one can only reply that production and consumption are connected to pleasure and resistance as well as domination, and that debates over health care are partially conducted through popular media. Cultural studies even has its own highly visible scandal(s). As some of you may recall, the cultural studies journal *Social Text* became mired in a public controversy over social constructionism and scientific truth claims in 1996, when a physicist published a paper there stating things he did not actually believe and then announced this fact in a populist academic magazine. He claimed that his hoax was evidence of sloppy thinking and its weakness as a site for radical politics. Given this deceitful conduct, it is no surprise that there is a bureau in the U.S. government dedicated to investigating federal research grant fraud.

So what is going on with these critiques? It seems as though cultural studies is an irritant to hegemonic forces because of its radical antielitism. This is antagonizing to both traditional academic disciplines and certain media mavens, who see the humanities' sacred duty as elevating the population (or at least segments of it) through indoctrination with a sacred array of knowledge carefully removed from the everyday that will "articulate . . . the essential elements of the national culture" to individuals and bind them to their societies through collective identification with ideas and practices (Aronowitz 2000, 4). Lawrence Grossberg (1997) explains the situation thus:

> On the one side, there is a discourse of multiculturalism and liberation which calls for a democratic culture based on a theory of identity and representation. On the other side, there is a discourse of conservatism based on canonical notions of general education and a desire to impose what it cannot justify—the existence of an illusory common culture. (381)

Public Law 89-209, which brought the Endowments for the Humanities and the Arts into being in 1965, modestly comments that "the world leadership which has come to the United States cannot rest solely upon superior power, wealth, and technology, but must be solidly founded upon the worldwide respect and admiration for the Nation's high qualities as a leader in the realm of ideas and of the spirit" (9). While the DeLillo quip about full professors reading cereal boxes is funny, it has a somewhat wry tone as well. It seems odd to turn away from high cultural pursuits and invest one's academic capital in the banal, to shift direction from Marx to fun parks and do so in the name of oppositional politics. The patented quip "Disneyland for the feebleminded" is also amusing. But in each case, there is something behind the remark that eludes its maker. Understanding the iconic significance and material history of American food is of great importance, while acknowledging the pleasures of ordinary people rather than privileging the quasi-sacerdotal pronouncements of an elect may not be so much "feebleminded" as threatening to cultural elites, who are committed to the formation of commitment via the study of literary classics (Aronowitz 2000, 4).

But while the pro–cultural studies British Quality Assurance Agency for Higher Education argues for the centrality of education in popular media criticism as a means of "mapping the contemporary," its measurement methods represent one more instance of imposing scientific management on intellectual workers. It is no surprise, then, that the British government's Tayloristic Research Assessment Exercise of 2001 saw the destruction of the Institute for Mass Communication Research at the University of Leicester, the founding site of British political economy of the media, and the Department of Sociology and Cultural Studies at the University of Birmingham, the founding site of British cultural studies.

There are certain ways in which there is nothing antielitist about cultural studies. First of all, lots of researchers in the discipline are academic stars. They are very well known, their books sometimes sell well, and they have comfortable research jobs. But many are at risk in adopting teaching strategies and research methods that put them at odds with occupational norms, leaving them out on a limb. In broader labor terms, cultural studies frequently do not work in the interest of academic labor. There are many stories that are told of, say, a French or Spanish department that has a medievalist position, or an English department with an Old Norse position. A couple of jobs then fall vacant (people retire, leave, or something else happens). There is pressure from the administration to diversify the department and save money, so a decision is made to make a multicultural hire and turn two jobs into one. A minority person hired to fill that job then faces onerous tenure requirements, difficult political constituencies to manage, and resentment on the part of the people in the original departments over the continuing loss of specialist positions. So cultural studies, in that sense, enables forms of control that disempower traditional faculty while putting minority faculty at risk. So where can cultural studies turn in this context of governmentalization and corporatization that will be, at the same time, pragmatic?

My proposal is cultural policy studies, because it can work in the context of these two processes without enabling them, while at the same time holding an appreciation of issues such as the labor strife, to which I have alluded.

The key appeal of cultural texts lies in their meanings. The conjunction of socioeconomic and representational analyses is the natural allies explaining them. But certain camps on both sides have maintained that they are mutually exclusive on the grounds that one approach is concerned with structures of economy and the other with structures of meaning. This need not be the case. Historically, the best critical political economy and cultural studies have worked through the imbrication of power and signification at all points on the cultural continuum.

Ideally, blending the two approaches would heal the divisions between fact and interpretation as well as between the social sciences and the humanities, under the sign of a principled approach to cultural democracy. To that end, Lawrence Grossberg recommends "politicizing theory and theorizing politics" for a combination of abstract and grounded analysis. This requires a focus on the contradictions of organizational structures, their articulations with everyday living and textuality, and their intrication with the polity and economy, refusing any bifurcation that opposes the study of production and consumption, or fails to address axes of social stratification (1997, 4–5, 9–10).

Cultural Policy Studies

What is the existing state of environment of English-language cultural policy studies? The ur-texts on the economics of cultural assistance was a work published twenty years ago by Australian-based researchers, and one written thirty years ago by Herbert Gans on "taste cultures," both of which provided multilayered frameworks for intellectualizing the popular and its relationship to policy (Throsby and Withers 1979; Gans 1974, 121–59). Cultural policy studies was named and undertaken in the 1970s through the formation of the Association of Cultural Economics and the Center for Urban Studies at the University of Akron. This was followed by regular conferences on economics, social theory, and the arts, and major studies of policy and program evaluation were produced at Canada's Institute for New Interpretive Creative Activities, the Cultural Policy Unit of the Johns Hopkins Center for Metropolitan Planning and Research, the Cultural Information and Research Centres Liaison in Europe, and Columbia University's Research Center for the Arts and Culture. Publications such as the *Journal of Arts Management, Law and Society*, and the *Journal of Cultural Economics* have long provided a wealth of theoretical speculation and empirical reporting, and later they established connections to Washington, D.C.'s, Center for Art and Culture.

These developments have led to queries about the relationship between the humanities and a sociological basis of this area: for instance, whether there should be a "social science aesthetic," questions regarding how the new proxemics of administration, politics, and the arts should be ethically and technically managed, along with the need for both "historical analysis in policy making" and "the history of policy." But, in general, the social sciences side to the analysis of culture has been holding on to valueless shibboleths and has not initiated any significant progressive social change.

By contrast, cultural studies has an overt political agenda concerning social movements and cultural workers' rights. Angela McRobbie calls cultural policy "the missing agenda" of cultural studies, given that it does offer a program for change (1996, 335). But Stuart Cunningham suggests that

> many people trained in cultural studies would see their primary role as being critical of the dominant political, economic and social order. When cultural theorists do turn to questions of policy, our command metaphors of resistance and opposition predispose us to view the policy making process as inevitably compromised, incomplete and inadequate, peopled with those inexpert and ungrounded in theory and history or those wielding gross forms of political power for short-term ends. These people are then called to the bar of an abstrusely formulated critical idealism. (1992, 9)

The notion that theory underlies practice via a renewing critique taken up by bureaucracies has often seemed misplaced in the cultural field, where everyday academic critical practice eschews such relationships as either insufficiently aesthetic or too co-optive. Cunningham attacks this line of argument as failing to acknowledge, for instance, that public action on sexism in advertising and the status of women in the workplace has come about because of a shift from utopic critique to implemented policy. He calls for cultural studies to adopt a "political vocation" that draws its energies and direction from "a social democratic view of citizenship and the trainings necessary to activate and motivate it."

This "new command metaphor" will displace "revolutionary rhetoric" with a "reformist vocation." Its "wellsprings of engagement with policy" can nevertheless avoid "a politics of the status quo—a sophomoric version of civics," because cultural studies' ongoing concern with power will always be grounded in radicalism. Cunningham uses cultural policy studies as a conduit to cultural rights, access to information held by multinational corporations and international organizations, the means to balance power between developed and developing countries, and how all these developments have an impact locally (1992, 11). Jim McGuigan welcomes this turn in cultural studies, providing that it retains its radical insights by then connecting this to the emphasis in political economy concerning public debate and citizenship rights (1996, 21).

Those who want to see the full exercise of democracy must beware of falling for the rhetoric of citizenship that has been adopted in discriminatory and exclusionary ways, and must require each part of the consumer-citizen divide to illustrate its account of consumption or citizenship; the relationship between capital, democracy, knowledge, and diversity; and the role of the state in consumption, and of corporations in citizenship. Lastly, we must look to minority, indigenous, and migrant interests any time we are told consumers are unmarked or that citizens lie at the center of culture within borders. This profound contextualization will equip faculty and students to comprehend governmentalization and corporatization and do something about them.

Following are Dr. Miller's answers to questions posed during a discussion of his paper at the University of Illinois Center for Advanced Study.

Q: With the critique of knowledge production that you have given us today, how would you answer the question of competing rhetoric within the local context; that is, how would you encourage the conditions under which students and faculty work in the College of Fine and Applied Arts or Theater, et cetera? And how can such fields compete with the rhetoric of science and technology?

A: *I think that there are two issues here. The first issue is that of figuring out how to sell oneself to the university community as being something or someone of value; that is, how can one sell oneself to those outside of it? The second issue concerns curriculum. Concerning curriculum, keep in mind that I have taught in a kind of conservatory, and most of my colleagues are set builders and designers, et cetera. And while for many of them it is hard to see the value of taking classes to satisfy social or natural science requirements, after graduating, these same students want to know things like how to write a grant. Why? Because simply expressing yourself is not very fashionable to most cultural policy administrators; expressing the social needs of the community is very attractive.*

Again, because we are working within a hypercommodified culture, you need to be able to show others (i.e., the local municipal government, the administrators of the university, et cetera) that your work and research can have an impact beyond simply the abstract notion of "icing on the cake" and can provide vital sustenance to underserved communities. And what this requires is for those of us in the field to be prepared and ready with an agenda that is not organized just in terms of occupational advancement and traditional professional norms, but is organized in terms of cultural norms and cultural debates and that examines connections between power and subjectivity.

Q: It seems that the liberal arts are still valuable to the changing university, not only pedagogically in terms of curriculum, but also in the fact that they help train a labor force for an economic system that is increasingly given to flexible accumulation. In that sense, are the liberal arts salient for or within the new economy?

A: *As far as I am concerned, it is perfectly reasonable to try and train people in the three kinds of citizenship—political, economic, and cultural—and then allow the contradictions that will emerge from the knowledge that one hopes they derive from that type of training to take whatever form they may. For example, all those undergraduate students who are downloading music or films outside the copyright market will eventually learn and understand the consequences of that type of activity in connection with court decisions, the nature of copyrights, and property rights, et cetera, especially given the opportunity to think about the politics of private property. This can lead to all kinds of very interesting and sophisticated politics. And although you do not know which way people will go, a liberal arts education gives people the tools to be able to think through what their means of existence are, so that they will be able to come up with a good set of critical reasons for and against various kinds of applied research training and the future of their professionalism.*

Q: What do you think could remedy the imbalance of funds that come into any given research university? Because, any way you look at it, the reality is that medical schools, technology departments, et cetera have a greater pool of funds at their disposal than a classics department. Is there a solution to a situation like this? Because it seems that the lack of funds could easily drive these smaller not-as-"applied" fields out of the university community.

A: *I think that is where many people's concerns lie, because rewards and financing that are made available in certain sectors of the university are creating a sense of internal competition within the university community. However, in another kind of world, a world where universities are internally collaborative institutions, a professor of medicine and a classics professor could sit down and discuss how to share the funding in order to reinforce the work being done in classics. Now this kind of exchange, sometimes called cross-subsidization, increasingly does not happen because of this sense of every unit for itself. And yet collaboration is just as much a part of the building up of the U.S.—"e pluribus unum"—out of many comes one—as it is about working together for one purpose and goal in a university, something that has definitely been lost as our culture continues to drive toward a market-driven ethos.*

Q: There are many who are concerned about the corporatization of the university because it could even potentially get rid of the tenure system. What kind of issues does this raise concerning intellectual freedom?

A: *That is a really interesting question, and in fact, you are raising a whole series of crucial issues. While every faculty member can think of at least one face that does not seem to "pull their weight" in the department, generally people would agree that working people have much to contribute if the conditions and context are right, and if people are prepared to be lifelong learners. I am a staunch believer in the tenure system, first of all, because I am committed to income security. However, with regard to intellectual freedom, there is a threat of corporatization: although we have freedom of speech in the U.S. because of the First Amendment, there has been increasingly concentrated media ownership and, consequently, niche markets. Yet the university remains the one forum for the discussion and debate of ideas at great length. Tenure, and the job security that comes with it, is essential in creating an ethos conducive for intellectual dialogue and debate.*

REFERENCES

Aronowitz, Stanley. 2000. *The knowledge factory: Dismantling the corporate university and creating true higher learning.* Boston: Beacon Press.

Aronowitz, Stanley. 2001. *The last good job in America: Work and education in the new global technoculture.* Lanham, MD: Rowman & Littlefield.

Barthes, Roland. 1973. *Mythologies.* Trans. by Annette Lavers. London: Paladin.

Bhagwati, Jagdish. 2002. Coping with antiglobalization: A trilogy of discontents. *Foreign Affairs* 81 (1): 2–7.

Blumenstyk, Goldie. 2002. Universities try to keep inventions from going "out the back door." *Chronicle of Higher Education,* May 17, A33–34.

Boren, Mark Edelman. 2001. *Student resistance: A history of the unruly subject.* New York: Routledge.

Chaker, Anne Marie. 2002. State schools plan big tuition jumps: Major universities to increase fees by up to 25% as budget woes mount. *Wall Street Journal,* June 20, D1.

Chomsky, Noam. 1965. *Aspects of the theory of syntax.* Cambridge, MA: MIT Press.

Coombes, Annie E. 1994. *Reinventing Africa: Museums, material culture and popular imagination in Late Victorian and Edwardian England.* New Haven, CT: Yale University Press.

Cox, Ana Marie. 2000. Study shows colleges' dependence on their part-time instructors. *Chronicle of Higher Education,* December 1, A12–A14.

Cunningham, Stuart. 1992. *Framing culture: Criticism and policy in Australia.* Sydney: Allen & Unwin.

Donzelot, Jacques. 1979. *The policing of families.* Trans. Robert Hurley. New York: Pantheon.

Dryzek, John S. 1994. *Discursive democracy: Politics, policy, and political science.* Cambridge, UK: Cambridge University Press.

Eco, Umberto. 1972. Towards a semiotic inquiry into the television message. Trans. Paola Splendore. *Working Papers in Cultural Studies* 3:103–21.

The Economist. 2000. Pupil power. November 18, 40.

Fogel, Aaron. 1993. The prose of populations and the magic of demography. *Western Humanities Review* 47 (4): 312–37.

Foucault, Michel. 1978. Politics and the study of discourse. Trans. A. M. Nazzaro. Rev. C. Gordon. *Ideology and Consciousness* 3:7–26.

Foucault, Michel. 1984. *The history of sexuality: An introduction.* Trans. Robert Hurley. Harmondsworth, UK: Penguin.

Foucault, Michel. 1991. Governmentality. Trans. Pasquale Pasquino. In *The Foucault effect: Studies in governmentality,* ed. Graham Burchell, Colin Gordon, and Peter Miller, 87–104. London: Harvester Wheatsheaf.

Foucault, Michel. 1994. Problematics: Excerpts from conversations. In *Crash: Nostalgia for the absence of cyberspace,* ed. Robert Reynolds and Thomas Zummer, 121–27. New York: Third Waxing Space.

Gans, Herbert J. 1974. *Popular culture and high culture: An analysis and evaluation of taste.* New York: Basic Books.

García Canclini, Néstor. 1995. *Hybrid cultures: Strategies for entering and leaving modernity.* Trans. Christopher L. Chiappari and Silvia L. López. Minneapolis: University of Minnesota Press.

Goldschmidt, Nancy P., and James H. Finkelstein. 2001. Academics on board. *Academe* 87 (5): 33–37.

Gopnik, Adam. 1994. Read all about it. *New Yorker* 70 (41): 84–102.

Grossberg, Lawrence. 1997. *Bringing it all back home: Essays on cultural studies.* Durham, NC: Duke University Press.

Lewontin, R. C. 1997. The cold war and the transformation of the academy. In *The cold war and the university: Toward an intellectual history of the postwar years,* ed. Noam Chomsky, 1–34. New York: New Press.

Lyotard, Jean-François. 1988. *La condition postmoderne: Rapport sur le savoir.* Paris: Les Éditions de Minuit.

McGuigan, Jim. 1996. *Culture and the public sphere.* London: Routledge.

McRobbie, Angela. 1996. All the world's a stage, screen or magazine: When culture is the logic of late capitalism. *Media, Culture and Society* 18 (3): 335–42.

Martin, Randy. 1998. Introduction: Education as national pedagogy. In *Chalk lines: The politics of work in the managed university,* ed. Randy Martin, 1–29. Durham, NC: Duke University Press.

Minogue, Kenneth. 1994. Philosophy. *Times Literary Supplement,* November 25, 27–28.

Morley, David. 1998. So-called cultural studies: Dead ends and reinvented wheels. *Cultural Studies* 12 (4): 476–97.

New York Times. 2000. Unions and Universities. November 25, A18.

Pietrykowski, Bruce. 2001. Information technology and commercialization of knowledge: Corporate universities and class dynamics in an era of technological restructuring. *Journal of Economic Issues* 35 (2): 299–306.

Poovey, Mary. 2001. The twenty-first century university and the market: What price economic viability? In *Differences* 12 (1): 1–17.

Postrel, Virginia. 1999. The pleasures of persuasion. *Wall Street Journal,* August 2.

Quality Assurance Agency for Higher Education. 2002. *Communication, media, film and cultural studies.* Gloucester: Quality Assurance Agency for Higher Education.

Rhoades, Gary, and Sheila Slaughter. 1998. Academic capitalism, managed professionals, and supply-side higher education. In *Chalk lines: The politics of work in the managed university,* ed. Randy Martin, 33–68. Durham, NC: Duke University Press.

Rhoades, Gary, and Barbara Sporn. 2002. Quality assurance in Europe and the U.S.: Professional and political economic framing of higher education policy. *Higher Education* 43 (3): 391–408.

Sellars, Richard West. 1997. *Preserving nature in the national parks: A history.* New Haven, CT: Yale University Press.

Shapiro, Michael J. 1993. *Reading "Adam Smith": Desire, history and value.* Newbury Park, CA: Sage Publications.

Simpson, Christopher. 1996. *Science of coercion: Communication research and psychological warfare, 1945–1960.* New York: Oxford University Press.

Sora, J. W. 2001. Let's pretend we're a corporation: An introduction to the academic/corporate convergence. *Corporate Governance* 1 (1): 39–45.

Stimpson, Catharine R. 2000. A dean's skepticism about a graduate student union. *Chronicle of Higher Education*, May 5, B7.

Streeter, Thomas. 1996. *Selling the air: A critique of the policy of commercial broadcasting in the United States.* Chicago: University of Chicago Press.

Throsby, C. David, and Glen A. Withers. 1979. *The economics of the performing arts.* New York: St. Martin's Press.

NOTES

1. The Carnegie classification system identifies 125 U.S. educational institutions as research schools. They are my primary focus here.

2. Robert Triffin acted as plenipotentiary for the United States to the European Economic Community and then as a European delegate to the International Monetary Fund, just a few months apart, in the 1980s.

3. For example, Project Camelot in the 1960s.

4. For example, the policy and program management of the National Parks has consistently owed much more to bureaucratic force majeure, tourism money, and "development" than to ecological science (Dryzek 1994, 117; Sellars 1997, 3–4).

13

Technology and the Humanities in the "Global" Economy

Masao Miyoshi

The university is undergoing a rapid and fundamental change. While everyone knows this, the extent of its transformation, the change in direction, and the altered role within a society (which is itself changing rapidly) are problems we will continue to puzzle over for some time to come. In the hopes that it may not yet be too late for concerned people to intervene, I would like to point out several crucial developments for further consideration.

Four events come to mind: the privatization of the university since the late 1970s; the intense acceleration of high-tech R & D, especially in information technology (IT) and biotechnology; the general decline of the humanities, social sciences, and even basic sciences; and a revolutionary change in university management. Before I conclude, I would like to suggest some thoughts concerning the future of the university, civilization, and our planet earth as an inclusive totality.

The privatization of the university became noticeable about the time the unrest of the 1960s began to subside. The U.S. economy was visibly shaken in the 1970s because of the exorbitant expenditures accrued during the Cold War. The floating dollar, the oil shock, and increasing competition from other industrial nations were enough to place on center stage the ideas of competition and survival. The Thatcher and Reagan policies of neoliberalism, which intensified the predominance of private property, market pricing, and profit motive, resulted in a preoccupation with productivity and efficiency. Public universities came under the scrutiny of state legislators and the public in this matter of efficiency and productivity, accountability and accounting. The practical utility of higher education was now expected—partly as a reaction to what was perceived as the abuse of public funds by "irresponsible" students and professors. Colleges and universities were faced with the pressure of downsizing and cutting costs, and the talk of privatizing some public universities (e.g., UC Berkeley and UCLA) became more than just passing rumors.

The bipartisan patent law, the Bayh-Dole Act of 1980, which allowed the university to retain patents for its federally funded inventions, was in retrospect an epoch-making change, at least as important in social consequences as legislations like the Morrill Land Grant Act of 1862 or the G.I. Bill, signed in 1944. The Bayh-Dole Act enabled a corporation to acquire exclusive rights to a patent that was generated by federally funded university research in exchange for the payment of fees and royalties to the universities. Since corporations need not invest in the initial R & D and can choose what appears most profitable, the Bayh-Dole Act provides them with great opportunities. On the other hand, it creates another avenue of income for the university, while its applied science faculty, now called "inventors," split the royalties (anywhere from 25 to 50 percent) with the university and can build an increasingly sizeable fortune to supplement their university salary.

The number of U.S. patents awarded to the one hundred leading U.S. research universities grew from 177 in 1974, to 408 in 1984, and to 3,661 in 1999, increasing about twenty times within a quarter of a century (Association of University Technology Management [AUTM] Licensing Survey, FY1999). In 1999, 3,914 new licenses and options were given, and the adjusted gross license income amounted to $862 million, compared to the $725 million in 1998. Since 1980, at least 2,922 new companies have been formed thanks to licenses given from an academic institution, including 344 that were formed in FY1999 alone. AUTM also reports that $40.9 billion and 271,000 jobs can be attributed to the results of academic licensing. Furthermore, the tax revenues from these activities are reportedly $5 billion at the federal, state, and local levels. These are, in themselves, impressive figures. However, when placed in the context of the $8 trillion U.S. economy, this is indeed a pittance. Even in terms of the public and private support of university research, which amounts to $26.8 billion, the return of $862 million (we are talking about gross, not net, income) is rather poor business. So, what is all the fuss about high-tech R & D in the university?

I believe that the reason for this lust after university-industry alliances is the underlying structure of the capital movement. The support for university research overwhelmingly originates in federal sources: $16.8 billion compared to the $2.7 billion from industry. That means taxpayers' money subsidizes university R & D. Yet the fees generated by such research benefit the universities, the faculty involved in R & D, and the corporations making profits with relatively low risks as they sell their products at the end to consumers, that is, the taxpayers. Technology transfer not only propels technology from the university to industry, but capital from the taxpayers to the aggregate of the university as an institution, a portion of the faculty, as well as IT, biomedical, and other high-tech corporations.

Of course, the university-industry alliance is not built on the trade of patents and fees alone. As the need for information and technology continues to mount, corporations are bringing their sites of operation closer to the universities, observable in the research parks that are dotting the country from Silicon Valley to Kendall Square, and from Research Triangle to Silicon Hills. A research park is the representative spectacle of the twenty-first-century university campus, just as Gothic architecture was in the nineteenth century. In addition to the creation of start-ups, industrial personnel are being retrained in the university labs, while graduate students and academic researchers seek employment in industry. Applied science professors can become billionaires, as in the case of the founders of Genetech or Qualcomm. And, as if to authenticate the genuineness of the collaboration, corporate endowments in universities have dramatically increased in the last decade. There are an increasing number of faculty members taking temporary leave to work for corporations. In the Business Administration Department of UCLA alone, about 20 percent of the professors are involved in start-ups; at Carnegie Mellon, an estimated 25 percent of computer science professors have interests in outside ventures; and at Stanford, ten out of the forty-five computer science professors were on leave last year.

While this is publicly beneficial because it does bridge the gap between the corporate world and the university, allowing the university to better train its graduates to be effective workers, we must at the same time raise the conflict-of-interest issue that threatens to taint research with the bias of self-interest. Nationally, corporate part-time faculty employees have increased from 21 percent in 1993 to 33 percent in 1999, according to the newsletter of the National Association of State Universities and Land Grant Colleges (June 2000). University trustees and administrators often also sit on corporate boards for sizable wages, and corporate managers conversely serve as university regents. Furthermore, the university president who increasingly calls him- or herself the "CEO" of a university now acts like a corporate CEO and expects his faculty to behave like corporate employees. The notorious gap in wages between the corporate CEO and the manufacturing worker in the United States, which amounted to about 475 to 1 in 2000, may very well be adopted by the wage structure of academic employment.[1] Corporate terms such as efficiency, productivity, consumers, products, downsizing, and name brands (Harvard, Stanford, etc.) are unselfconsciously deployed. Together with these terms, corporate concepts of administration and management are being appropriated. Thus, the university president is more often than not a corporate manager and not an educator or scholar by training.

Knowledge and information are now being redefined as exchange commodities. Use and utility, then, determine the value of knowledge; that is, we are seeing market forces driving the direction of higher education. Learning is being radically transformed into intellectual property, the exclusive accumulation of which is now the highest objective of any research university. Consequently, many public research universities are now virtually private universities, whose state funding is as low as 20 percent, as in the case of the University of California, or like the University of Michigan, where less than 10 percent of the total operating budget and less than 20 percent of the general and education budget is allocated to the university by the state government. Thus, the chancellor of the University of California at San Diego (UCSD), who was a researcher manager at Bell Labs before joining the UCSD faculty as a physicist, recently declared in his state of the campus address: "As scholars, we should not seek knowledge for its own sake. We should seek knowledge that has real value for us and for our community." While the search for knowledge indeed should not be clouded by personal self-interest, his definition of the "real value" of university research clearly misses the discussion of who the constituents of "our community" are.

In connection to this, I must question the problems surrounding the conflict of interest and commitment. By now, everyone has his or her choice of episodes: Dr. David Kern's dispute with Brown University, Dr. Betty J. Dong's fight with her funding source about the disclosure of her findings, or Peter Taborsky's astonishing fight with Dean Robert Carnahan at the University of South Florida, ending with Taborsky being sent to a chain gang at the state penitentiary's maximum security unit.[2] Dr. Sheldon Krimsky's studies of the relationships between the researchers writing articles published in leading medical journals and their invested financial interests are known to anyone familiar with the collaboration of business and academia. Furthermore, many research university administrators are involved in the adjudication of conflicts among faculty, ex-faculty, and graduate students, as well as affiliated start-ups, in efforts to settle them quietly out of court while protecting the reputation of the university, another reason why a good number of university technology transfer offices are losing money despite their gross incomes.

The sudden spread of conflict-of-interest cases is alarming enough. However, a more serious issue is that of a new definition being imposed on the idea of "conflict of interest." The perception that there is a "conflict of interest" does not in itself constitute a crime, but "conflict of interest" is simply a *perceived situation* in which an involved party has vested personal interest that is *likely* or even *possible* to interfere with required neutrality. It is always a matter of the presence of a *potential* of improper compromise. Once a conflict of interest is materialized in an act that wrongfully favors the interest of the party in question, that conflict of interest is already a criminal act—theft, perjury, insider trading, or whatever—punishable by law. As Krimsky discovered in his investigation, however, many editors of leading journals now insist that a conflict of interest does not exist until the wrongdoing is proven; thus, like a criminal, one should be presumed innocent until proven guilty. Even the disclosure of financial interests is dismissed as unneeded. This is non-sense as well as dangerous. A conflict of interest is always based on *perceived* interest, since the "actual" conflict is logically impossible to verify unless it has already turned into a crime as argued above. Editors of journals such as *Lancet, Nature,* and others, however, have attacked Krimsky's effort, calling it the "new McCarthyism in science."

Conflicts of commitments may be even more widely spread, and the trans-gression is just as hard to prove. Faculty members are typically tied to the campus, typically for four days a week during nine months of the year. But for the rest of the time, they are free to invest their time in whatever private endeavors they choose to undertake. The question is, however, whether time can be so clearly divisible and discrete. Is the line between intellectual research and commercial enterprise self-evident? For example, in addition to the widely spread aforementioned part-time practice, there are cases like the LECG (the Legal and Economic Counseling Group), a firm formed by four active faculty members at UC Berkeley, specializing in legal and economic consulting. The earnings of the members of the firm ranged from $14 million in LECG stock after the initial public offering, to $33 million in stock. (Note that these figures are at least four years old, before the bubble burst; their cur-rent stock holdings are not available.) One of the law professors in the group has been a senior economist on the Council of Economic Advisors, and another major shareholder is currently on leave and serving as the deputy assistant attorney general of antitrust at the Justice Department. Laura D'An-drea Tyson, the Bank of America Dean of the Haas School of Business, the former chair of the White House Council of Economic Advisors, and also for-mer chair of the National Economic Council, is also another member of this particular firm specializing in antitrust, environmental and natural resources economics, intellectual property, and other fields. LECG's clients not only include large corporations, but also whole governments such as Argentina, Japan, New Zealand, and others.[3]

There are about one hundred private and public universities in the United States engaged in technology transfer. The institution receiving the largest funding for sponsored research in 1999 was the University of California (UC) system, at $1.9 billion, followed by Johns Hopkins, Massachusetts Institute of Technology (MIT), University of Michigan, University of Wash-ington, University of Pennsylvania, University of Wisconsin at Madison, University of Minnesota, Stanford, and North Carolina State. However, the order is considerably different in terms of the size of the gross income in tech transfer: Columbia, $89 million; followed by the UC system ($74 mil-lion); then Florida State, Yale, University of Washington, Stanford, Michigan State, University of Florida, University of Wisconsin at Madison, and MIT. Although the ranking order changes from year to year, there is one constant aspect in technology transfer: there is no clear difference between public and private universities. They are both in dead pursuit of what can be generated in R & D activities.

In a book called *Academic Capitalism*,[4] its coauthors, Sheila Slaughter and Larry Leslie, posit that the university might become divided between commercially engaged and unengaged faculty. The authors predict that

> faculty not participating in academic capitalism will become teachers rather than teacher-researchers, work on rolling contracts rather than having tenure, and will have less to say in terms of the curriculum or the direction of research universities. (211)

While this eventuality has not quite taken place yet, the divide is already widening. For example, I have read a number of books discussing the twenty-first-century research university, and hardly any of them mention the humanities except a few times in the form of an aside. The R & D faculty are increasingly indifferent to the arts and the humanities, the social sciences, or even pure sciences (which the president of the University of California and others refer to as "curiosity research"), since they are the ones bringing money to the campus. Most grants have a 50 percent cost sharing, or indirect cost recovery, and many campus administrators have a discretionary power over a sizeable portion of these funds to promote more revenues. On the other hand, professors in the humanities are distant from and disdainful of the applied sciences, as if they share no common ground whatsoever. What is significant is how the humanities faculty have withdrawn into themselves. And the reasons they give for this retreat may only be a part of self-destructive defeatism that may not have any relationship to the razzle-dazzle of technology.

I will not go into the history behind the increasing isolation of the humanities from the "rest of the world" and its near complete neglect by those outside the profession.[5] Briefly put, the oppositionism of the 1960s and 1970s, engaged in anticolonialism, anti-imperialism, antiracism, and antipatriarchy, was a powerful libratory movement. Its ascendancy in the academy was a moment of victory celebrated in the name of justice and equality. It sought to challenge the unquestioned acceptance of enlightenment reason and universalism. However, in the past thirty years, multiculturalism has become naturalized and assimilated, gradually hardening to a new academic orthodoxy. The logic of difference now dominates. Since difference insists on the incommensurability of each group, any definable particular insists on its autonomy, like the neoliberal practice. It promotes exclusive self-interest. Combined with the traditional conservatism of the academy, the humanities is now a plurality of incommensurable autonomies. The fragmentation of the humanities among ethnic studies, gender studies, postcolonial studies, cultural studies, and conventional disciplines, all based on the principle of difference and particularism, is so intensified that its members are no longer sharing or even speaking with each other. All the while, technology continues to attract the attention of the public and the media, and the humanities appear like contentious groups of eccentrics, each speaking in exclusive and inaccessible codes, which sound like gibberish to the outsiders.

Fragmentation and atomization are taking place in nearly every discipline within the humanities. Yet, just as national borders are less meaningful for the corporate and academic practitioners than before, disciplinary borders are also losing their significance for their practitioners. In the humanities, from a single national language and literature to general literatures, for example, from eighteenth-century French history to some aspects of gender inequity, many younger scholars are restlessly looking away from the specific discipline in which they received their doctorates. Graduate students and assistant professors, however, are forced to become members of single disciplinary fields and departments, and for the moment, they acquiesce. As cultural studies begins replacing literary, historical, or sociological studies, however, crossing borders is becoming an intellectual compulsion for inquiring and critical minds. In fact, books dealing with received disciplinary topics are no longer being read or even published with enthusiasm, while transdisciplinary publications are being more eagerly produced—though the number of books being bought, read, or discussed, too, seems to be generally decreasing as well. Moreover, the traditional department structure works against the interdisciplinary program, center, and institute structure that would encourage rigorous integrated studies. Thus, nearly all research universities seem to encourage interdisciplinary projects in one form or another, yet these projects finally fail, simply because the faculty are still inescapably locked in a tradition-oriented department. Professors visit a program, center, and institute for a joint program with scholars from other disciplines, but at the end of the day, they all go home to their departments, to be tenured, promoted, and honored in their professions. As long as the present regime prevails, true transdisciplinarity is a dream that will remain a dream—except among those who are so secure in their profession that they can afford to cross borders and challenge their assigned work locations and job descriptions.

Unfortunately, established scholars do not always take the risk by venturing out into unfamiliar territories as younger scholars might. Their expected task is usually to secure a boundary for perpetuity and to be rewarded for their guardianship. Notably, even young scholars grow to reach that stage rather surprisingly fast. Thus, interdisciplinarity remains, not a collaborative work of interconnected disciplines, but a transitory site for isolated, closeted, and recalcitrant scholars, who visit for a brief time for pleasure but have to return to their home base for security.

This is particularly unfortunate when the university is being powerfully tempted by a "financialized" and monetized vision of itself, and in its heated pursuit of money and grandeur is ready to jettison itself to become uncritical and trivialized. With no persuasive input emanating from humanities scholars, administrators are not concerned with supporting these hapless disciplines and professions. The way things are currently going, the humanities departments in the research university will most likely turn into a service sector dedicated to teaching undergraduates how to read, write, and be tasteful. A slightly more serious role may be commissioned at the liberal arts colleges, if they can attract with their cherished brand names students still interested in general education.

A few words on distance learning. Obviously, online courses can travel, and thus a curriculum consisting of lectures by star professors from prestigious colleges like Oxford, Paris, or Berlin may invite students. In other words, brand names will appeal to upscale customers across the world. Accessibility anywhere and anytime is also a democratic instrument and can serve the world public. At the same time, the infrastructure for interaction—that is, Q & A, evaluation, certification, comments, and fraternization—must be constructed with greater care. Such a task will, however, nullify the raison d'être of the virtual university, since satisfactory interaction requires time and energy nearly identical to that of campus learning. In the meantime, the entire enrollment (including nonstudents) at virtual universities in the United States remains only 1.6 million, just about 10 percent of the 15 million students enrolled in higher education now. Downsizing faculty through distance learning remains, at least for now, simply wishful thinking of university managers.

Let me take two opposite theoretical possibilities vis-à-vis the academia-industry relationship: One extreme might be the merger of the university and the corporation, and the other extreme might be the Luddite retroversion to an imaginary golden age. A merger between the two could be possible. We have an example in the University of Phoenix, whose president was recently promoted to the CEO of the Apollo Group, its proprietary company, whose stock is currently doing exceptionally well. Mainstream university administrators are seriously examining the University of Phoenix model. A few faculty members devise the curriculum, and thousands of low-cost facilitators (unemployed PhDs and ABDs) from across the country then transmit that information to a student body. This practice is becoming more and more acceptable even among accredited, private American colleges and universities. Its success, high profit and large enrollment, as well as reputation among employers are, however, due to its practical and vocational orientation, which is not designed to stimulate intellectual curiosity or critical investigation. It might conceivably grow in sophistication some day, but as of now, that possibility seems to be very slim. However, corporate managers are not fond of academics, nor are they familiar with them, stereotyped as being too arrogant, distant, and long-winded. The merger of the two institutions seems as impossible as does the full maturation of the University of Phoenix.

A Luddite-like retroversion is perhaps just as implausible. Once the flood-gates are opened, it is nearly impossible to shut off the ensuing flow. The coalition of R & D scholars and commercial developers will resist and oppose with all their might, along with the support they can raise from various sectors, that is, corporate, political, financial, media, and even academic. Most students will also join them. Besides, the retroversion will not resuscitate general education, social sciences, and the humanities, all of which are in grave ill health for reasons unrelated to corporatization.

At the same time, the prevailing symbiosis of the small number of universities and corporate sectors of society is not acceptable to anyone engaged in the critical examination of the world. What, then, are the possibilities of collaboration in the space that lies between the merger and the Luddite resistance? The proposal must have some appeal to those who entertain nothing but self-interest. There are two core ideas that the public might pay attention to, if they are presented with reason and conviction. First, the public needs a body of informed critics making observations and recommendations regarding the direction of civilization as it charges ahead with the engine of technology. Of course, I do not mean that we in the humanities are adequately trained for this job or can claim spiritual authority on these matters. But those in the humanities can try to restore their reputation and qualifications by retraining their historical perspective and general knowledge of the world and by redirecting themselves seriously to assume this responsibility.

In fact, higher education as a whole is currently experiencing a nearly complete loss of its historic purpose. The concept of culture, we remember, was constructed in the nineteenth century as the nation-state was formed with a secular foundation. As Mathew Arnold most bluntly articulated, culture is in service for the "state." As nation-states were replaced later by transnational corporations as the driving engine of the world political economy, the idea of culture rapidly lost its significance. The idea of "excellence" that is substituted nowadays as the objective of education is contentless.[6] It means nothing more than "success," just as the logo of Hofstra University unembarrassedly declares (we "teach success") in their regular advertisements in the *New York Times* and the *New Yorker*. Excellence in what, or success in what, is never mentioned. And yet one does feel that Hofstra is at least more honest in its vacuous advertisement than excellence-promoting institutions, or even those traditional schools that were devoted to the interests of the nation-state, while claiming academic freedom and autonomy.

One all-encompassing purpose of education can be found in teaching the interconnectedness of "nature" in its entirety. That is, the totality of the planet, all human beings in every place on earth, all living creatures and their relations, and all materials that constitute the earth, which ultimately we all share. I am not just referring to the environmental sustainability or ecological survival of species, but to the very planet we all share as one commonality. Of course, there are many determined and powerful corporate and political leaders who turn a blind eye toward the hazards of climate change; the depletion of natural resources; air, water, and land contamination; ozone-layer reduction; vanishing biodiversity; and many other crises around us. The rich will try to stay as far away as possible from the deterioration, but even they cannot remain protected for long. Such stubborn ignorance will not and cannot last. The Kyoto Protocol is far from satisfactory, but it is at least the minimal first step, with a majority of people accepting the minimal ceiling of the carbon dioxide production level. Sooner or later, even the most recalcitrant, like the United States, will have to realize that our current state of capitalism is consuming far beyond the planet's capacity, and that their children and children's children will have to face the global catastrophe they do not deserve.

Once we accept this idea of a planet-based totality, we might for once be able to agree on how to devise a way to share our true public space and resources with the rest of the world. No single group can shape the future of the planet; the task must be executed as a collaborative effort. We must begin seriously investigating our consumption, which is extremely imbalanced, as discussed earlier. The uneven distribution of wealth means nothing but a pattern of uneven consumption and production. In order to assure the future of the planet, the consumption of natural resources of the last quarter century (which equals the sum of all consumption from the beginning of recorded human history) must be severely curbed. In industrialized societies, especially the United States, unrestrained consumption can only be described as criminal when one billion people on earth, who live on less than one dollar a day, desperately need to increase their consumption for survival. If the people of China and India decide to own automobiles at the same rate as Americans, the planet's atmosphere could not possibly be sustained. And yet the automobile manufacturers of developed countries are competing with each other to reach these two billion future buyers.

The university can and needs to reconsider its research in the light of this commonality of a shared planet. It is not completely clear what form this research on the future of the planet will take. However, it must combine environmental engineering with economics, political science, and cultural studies so that scholars in all fields may be able to develop a political economy and culture aiming to reduce consumption without cutting employment. The reduction of waste in the industrial world must be simultaneously carried out with the increase of production and consumption in many of the developing nations. They need to devise ways to train and integrate the unused labor forces of the Third World to equalize wealth. Some questions we can ask, then, include, Can runaway capitalism be restrained? In order to develop a nongrowth capitalism, what mechanism needs to be invented? Who owns the mode of production now? How does production relate to consumption?

What does finance capitalism consume? Can reduced-wage labor be redirected to a richer quality of life? By far, the most difficult task in this project is how to persuade people that culturally and politically there is no other future for any of us. Whether through publications, schools, universities, NGOs, UN-affiliated organizations, the media, or other residues of the state apparatus, we must somehow communicate the need to curb our material dream and work toward an agreement concerning our unavoidable future. For such a future, we need to reconceptualize a universally shared culture, as we have never done in human history.

We may very well fail in this endeavor, too. But if we do, we will not be there to see it. Perhaps, in fact, we deserve to perish. On the other hand, faced with a universally inescapable fate of destruction and nullification, we may yet find a way to confront it together, while learning to coexist with others. There is at least that promise of hope, the only hope we have been allowed to entertain together with everybody else on this planet.

Following are Dr. Miyoshi's answers to questions posed during a discussion of his paper at the University of Illinois Center for Advanced Study.

Q: Your talk ends with the discussion of a rudimentary research plan, in some respects, for the humanities. Given the first part of your discussion about the climate of universities, do you see this plan as feasible, given the situation of the university that you have outlined? Because you are essentially calling on humanities people to start doing research critical of the sources of funding now increasingly funding the university, not to mention the states that are or are not funding the universities, et cetera. I was wondering if you could address the practical aspect of this sort of research.

A: *One of the reasons I place so much trust in this "plan" is my assumption that most people do not really know anything about the finiteness of the resources. This is true almost everywhere. For example, there is a very recent book called* Consumer Society Reader, *which is a rather good collection of essays. They talk about consumerism from Frankfurt School up to now. None of these thirty essays mentions anything about the finality of the exhaustion of natural resources. Lawrence Summers, president of Harvard, who was a senior researcher at World Bank, and the secretary of the treasury under Clinton, talked about the transfer of pollution and the toxic materials from overpolluted places like the U.S. to underpolluted places in the Arctic. Even he does not discuss the finiteness of natural resources, seemingly assuming that natural resources are infinite. Production means of course the extraction of resources and then transforming most of them into unrenewable waste. The negative aspect of this is not fully considered. Thus, the task of redefining production will be a huge paradigm shift that can be undertaken by the humanities as a discipline. If we succeed in establishing something like transdisciplinary studies, we can then talk about what the cultural studies of literature or culture can do to promote that idea so that we can reimagine and re-create the cohesiveness of society, which we seem to have completely lost.*

Q: It seems that one of the basic presumptions underlying your thoughts is the separation of the humanities and the sciences, which leads to a C. P. Snow-ish idea of the two cultures. The experience, however, in the humanities, could have changed. My observation is that if you talk to geneticists, for instance, as soon as the question of life comes up, they do not have any qualification in even relating that question to the treasure and the tradition of the humanities as far as this question is concerned. I do not know of any geneticist who is able to explain what it means to be a member of the family. In my own experience, during the last fifteen years teaching medical students, this has affected many hundreds of physicians in their view in relating to their practice, their immediate scientific practice, and output, even changing some PhD dissertations that were produced. Is not the challenge of today to the humanities, then, to better understand about what is happening in the sciences and to actually be in the sciences as sociologists, psychologists, or even as people specializing in French literature, because that would enrich the view of the scientists and make the social consumption of science somewhat different?

A: *I will answer the comment concerning C. P. Snow first. When C. P. Snow talked about two cultures in the 1960s, culture did exist. It was not related to science, but the idea of culture was there. Now I do not think we have anything like it. Culture has thoroughly transformed into some kind of commercial activity—consumerism, entertainment, tourism, et cetera. And so, what we pretend to have in the form of museum, art, music, are now instruments of commercial activities. It is very difficult to think about how to relate so-called cultural or artistic activities with what goes on in the name of science. Now, you talk about scientists being not familiar with the humanities, the notion of life, and so on, but the converse also prevails, that almost no one in the humanities pays any attention to things like genomics either. So the dilemma of this fragmentation suggests that somehow we have to build a coherent whole, and I am just proposing one way that this could be done. Certainly I do not have an answer, but one interesting thing within the humanities is the notion of literature, history, and all these disciplines seem to be really blurring the distinctions. And in that "fuzziness" can be found some hope for an integrative knowledge.*

NOTES

1. I have no figure for 2001, although an article in the *New York Times* by David Leonhardt ["For the Boss, Happy Days Are Still Here"], on April 1, 2001, announces that the gap has again increased during the last twelve months.
2. See Seth Shulman, *Owning the Future* (Boston: Houghton Mifflin, 1999), 106.

3. I will not go into the UC Berkeley College of Natural Resources and Novartis deal, since the case was outlined by Eyal Press and Jennifer Washburn in their widely read *Atlantic Monthly* article "The Kept University" (March 2000). All I can say at this point is that it would seem to take superhuman self-control to keep one's teaching and scholarly duties in balance, if one were to be engaged in such razzle-dazzle activities offering easy and ready millions.

4. Sheila Slaughter and Larry Leslie, *Academic Capitalism* (Baltimore: Johns Hopkins University Press, 1998).

5. See "Ivory Tower in Escrow," *Boundary 2* 27, no. 1: 7–50.

6. Bill Readings, *The University in Ruins* (Cambridge, MA: Harvard University Press, 1994).

Index

About the Editors

William T. Greenough is Swanlund Chair and Center for Advanced Study Professor of Psychology, Psychiatry, and Cell and Structural Biology at the University of Illinois in Urbana-Champaign. He is among the world's leading investigators of experience-related neuronal plasticity in the mammalian brain, and he has helped establish the idea that learning and memory involve the rapid formation of new synaptic connections between neurons as well as modifications of preexisting connections. Throughout his distinguished career, Professor Greenough has sought to understand the brain mechanisms underlying learning and memory. He has been a major proponent of the hypothesis that a key element in both development and memory in mammals is the sculpting of synaptic connections between neurons. More recently, his work has explored plasticity in nonneuronal cells of the brain as well as the cell biology of fragile X syndrome and alcohol-related neurodevelopmental disorders. Professor Greenough is a member of the National Academy of Sciences, a fellow of the American Psychological Society and recipient of its William James Fellow Award, a fellow of the American Psychological Association and recipient of its Distinguished Scientific Contribution Award, a member of the Society of Experimental Psychologists, and a University Scholar of the University of Illinois. He has received the Oakley Kunde award for undergraduate teaching and, in April 2003, the Award for Distinguished Scientific Contribution for the Society for Research in Child Development.

Philip J. McConnaughay is dean and the Donald J. Farage Professor of Law at The Pennsylvania State University Dickinson School of Law. Previously, he was a professor of law at the University of Illinois College of Law. Dean McConnaughay is the author of several scholarly articles and edited books concerning international commercial dispute resolution, the regulation of international commerce, and the role of arbitration in economic development. He has been a visiting professor at Northwest University in Xi'an, China, and he has lectured on development and intellectual property issues in Vietnam, China, and Europe. For eighteen years, Dean McConnaughay practiced law with the international law firm of Morrison & Foerster, including almost ten years as a resident partner in Tokyo and Hong Kong. His work included representing Fujitsu Limited in the celebrated IBM/Fujitsu Arbitration, a multibillion-dollar dispute concerning worldwide intellectual property rights in mainframe computer operating system software. Dean McConnaughay also served as an adviser to a Government of Indonesia project with respect to the drafting of a new arbitration law, and he consulted with the U.S. Department of Justice concerning its antitrust prosecution of Microsoft Corporation.

Jay P. Kesan is professor and director of the Program in Intellectual Property and Technology Law at the University of Illinois at Urbana-Champaign. A PhD in electrical and computer engineering from the University of Texas and former research scientist at the IBM T. J. Watson Research Center in New York, Professor Kesan also holds appointments with the University of Illinois Department of Electrical and Computer Engineering and the Institute of Government and Public Affairs. He has been a JSPS Invited Fellow and Visiting Professor at the University of Tokyo, a visiting assistant professor at Georgetown University Law Center, and the Distinguished Jerold Hosier Chair in Intellectual Property at DePaul University. Professor Kesan has written and published extensively in the areas of patent law and patent institutions, law and the regulation of cyberspace, intellectual property, and law and economics. His research on how information technologies regulate behavior is funded by the National Science Foundation, and his research on intellectual property protection for agricultural biotechnology is funded by the U.S. Department of Agriculture.

About the Contributors

Erich Bloch is a principal of the Washington Advisory Group, a distinguished fellow at the Council on Competitiveness, a member of the President's Council of Advisors on Science and Technology, a former director of the National Science Foundation (NSF), and the former corporate vice president for technical personnel development at IBM. While at IBM, Dr. Bloch was a key figure responsible for IBM's STRETCH Computer Systems Engineering Project and for the groundbreaking developments of the IBM 360 mainframe computer. His developments revolutionized the computer industry and led to his receipt of the National Medal of Technology in 1985. In his six-year tenure at the NSF (1984 to 1990), Dr. Bloch supervised the foundation's $3 billion annual budget and transitioned the NSFNET into a commercialized Internet, which has had a tremendous economic and societal impact from the 1990s to today. Dr. Bloch also established the NSF's Computer and Information Science and Engineering Directorate. As a distinguished fellow at the Council of Competitiveness, Dr. Bloch promotes policies advocating the use of innovation in the development of the U.S. economy. He was awarded the 1999 Robert N. Noyce Award by the Semiconductor Industry Association, its highest honor for leadership. Mr. Bloch was awarded the Vannevar Bush Award in 2002 by the NSF, which honors long-standing scientific achievement and outstanding contributions to the nation and humankind. Dr. Bloch is a member of the U.S. National Academy of Engineering, a member of the Swedish Academy of Engineering Sciences, a fellow of the Institute of Electrical and Electronics Engineers, and a foreign member of the Engineering Academy of Japan. Dr. Bloch's chapter is "The Changing Nature of Innovation in the United States."

Ann B. Carlson is Senior Staff Associate for Policy and Planning at the National Science Foundation.

Kelly A. Cassaday is the former head of communications at The International Maize and Wheat Improvement Center.

Lord Meghnad Desai is a world-renowned development economist and the founder and director of the Centre for the Study of Global Governance at the London School of Economics, where he also is a distinguished professor of economics. Lord Desai is the author of numerous books on economics, including *Marxian Economics, Applied Econometrics, India's Triple By-Pass,* and *Marx's Revenge: The Resurgence of Capitalism and the Death of State Socialism,* as well as two collections of essays, *Macroeconomics and Monetary Theory* and *Poverty, Famine and Economic Development,* and he is editor of *LSE on Equality, Global Governance,* and *China in a Changing World Economy.* Lord Desai was educated at the University of Bombay and received his PhD from the University of Pennsylvania. From 1984 to 1991, he was coeditor of the *Journal of Applied Econometrics.* He has been both chair and president of Isligton South and Pinsbury Constituency Labour Party in London and was made a peer in April 1991. He is currently chairman of the board of trustees for Training for Life, chairman of the management board of City Roads, and on the board of *Tribune* magazine. Lord Desai's chapter is "Can Universities Survive the Global Knowledge Revolution?"

Rebecca Eisenberg is the Robert and Barbara Luciano Professor of Law at the University of Michigan Law School. She has written and lectured extensively about patent law as applied to biotechnology and the role of intellectual property at the public-private divide in research science. Professor Eisenberg has received grants from the program on Ethical, Legal, and Social Implications of the Human Genome Project from the U.S. Department of Energy Office of Biological and Environmental Research for her work on private appropriation and public dissemination of DNA sequence information. She has played an active role in public policy debates concerning the role of intellectual property in biomedical research. She is a member of the Advisory Committee to the Director of the National Institutes of Health; the Panel on Science, Technology, and Law of the National Academies; and the board of directors of the Stem Cell Genomics and Therapeutics Network in Canada. Professor Eisenberg's chapter is "The Public-Private Divide in Genomics."

John H. Gibbons is internationally renowned for his contributions to physics, energy, environment, and technology and public policy. He served as assistant to the president for science and technology and as director of the Office of Science and Technology Policy from 1993 to 1998 under President Clinton. Prior to his White House service, Dr. Gibbons was the director of the U.S. Congressional Office of Technology Assessment (OTA). The OTA was a bipartisan, bicameral agency designed to serve congressional committees as their principal source of independent, expert, and comprehensive analysis on issues involving the impacts of science and technology on society. After leaving the White House, Dr. Gibbons served as the Karl T. Compton Lecturer at MIT from 1998 to 1999 and as senior fellow at the National Academy of Engineering from 1999 to 2000, where he assisted NAE's president on a variety of topics including the new NAE program in earth systems engineering. He was the senior adviser to the U.S. Department of State from 1999 to 2001, where he assisted the secretary in revitalizing science and technology capabilities, including creating the position of science adviser to the secretary. Dr. Gibbons also serves on a number of prestigious boards and committees in both the private and public sectors and is a member of such organizations as the American Association for the Advancement of Science, the American Physical Society, and the American Philosophical Society, to name only a few. Dr. Gibbons has received six honorary doctorates as well as many distinguished prizes and awards from world-eminent organizations. His chapter is "The University of the Twenty-first Century: Artifact, Sea Anchor, or Pathfinder?"

Michael K. Hansen is a research associate with the Consumer Policy Institute, a division of Consumers Union (publisher of *Consumer Reports* magazine), where he works on biotechnology issues. Dr. Hansen is the author of *Biotechnology and Milk: Benefit or Threat?* and has written reports for the Consumer Policy Institute and Consumers Union on household pest control, alternatives to agricultural pesticides in developing countries, and the pesticide and agriculture policies of the World Bank and U.N. Food and Agriculture Organization. Dr. Hansen has been largely responsible for developing Consumer Union's positions on safety, testing, and labeling of genetically engineered food. He has testified at hearings in the United States and Canada and has prepared comments on various proposed U.S. governmental rules and regulations on biotechnology issues. He has been quoted widely in national media on safety of genetically engineered foods. Dr. Hansen served on a biotechnology advisory committee set up to advise the Consultative Group on International Agricultural Research on biotechnology issues for the public sector International Agricultural Research Centers. He has served as an international expert in 1998 to the Food and Agriculture Organization/World Health Organization Joint Expert Committee on Agricultural Biotechnology. During 1999 and 2000, he traveled extensively in South Asia, Europe, and the United States as an invited expert on potential human health and environmental concerns related to genetically engineered crops. His chapter is "The Effects of University-Corporate Relations on Biotechnology Research."

Donald N. Langenberg, a groundbreaking physicist and nationally recognized advocate for higher education, is vice chairman of the National Council for Science and the Environment, chancellor emeritus of the thirteen-member University System of Maryland, and a professor of electrical engineering at the University of Maryland at College Park. He is a former acting director and deputy director of the National Science Foundation, having been appointed by President Carter in 1980. In 1983, he became chancellor of the University of Illinois at Chicago, where he was also a professor of physics. He is the recipient of the John Price Wetherill Medal of the Franklin Institute, the Distinguished Contribution to Research Administration Award of the Society of Research Administrators, and many other awards. Dr. Langenberg is the author or coauthor of over one hundred articles and has edited several books. He has held predoctoral and postdoctoral fellowships from the National Science Foundation, the Alfred P. Sloan Foundation, and the John Simon Guggenheim Foundation. He has had visiting teaching positions at Oxford, the Ecole Normale Superieure, the California Institute of Technology, and the Technische Universitat Munchen. Dr. Langenberg serves on a number of boards including the American Association for the Advancement of Science, the Alfred P. Sloan Foundation, the Board of Trustees of the University of Pennsylvania, and the Executive Board of the National Association of the State Universities and Land Grant Colleges. He is president of the American Physical Society and sits on the board of the National Council for Science and the Environment. Dr. Langenberg's chapter is "Research Universities in the Third Millennium: Genius with Character."

Toby Miller is professor of cinema studies at New York University and the codirector of the graduate studies program there. His work in popular culture looks at how television, magazines, gender, film, and museums influence the ways that our society thinks about itself and such issues as citizenship and democracy. Professor Miller is a participant in the Privatization of Culture Project, a collaborative study between New York City's American Studies Program, the Sociology Program at the New School for Social Research, and the Center for Cultural Studies at the Graduate Center for the City University of New York. Dr. Miller is affiliated with organizations such as the Modern Language Association, the American Film Institute, the Society for Cinema Studies, the Museum of Television and Radio, and Media Information Australia and *Social Text*. His professional expertise has been acknowledged by the Paulette Goddard Junior Faculty Fellowship at NYU and the Henry Mayer Memorial Essay Prize by the Center for International Research on Communication and Information Technologies. Selected published works include *Espionage on Film and Television* and *The Avengers*, and he is coauthor of *Contemporary Australian Television* and coeditor of

Screening Cultural Studies. His chapter is "The Governmentalization and Corporatization of Research."

Masao Miyoshi is the Hajime Mori Professor of Japanese, English, and Comparative Literature at the University of California at San Diego, and he is considered one of the most distinguished and widely cited authors and thinkers about the place of the humanities in the era of economic globalization, business administration, and biotech. Professor Miyoshi earned his PhD from New York University and published his first book, *The Divided Self: A Perspective on the Literature of the Victorians*, in 1969. His interests since have ranged from Victorian literature to modern Japanese literature, including English and comparative literature, cultural studies, and critical theory. In the past decade, he has emerged as a formidable critic of postmodernism and globalization, having edited, with H. D. Harootunian, *Postmodernism and Japan* in 1989 and coauthored *The Culture of Globalization* with Frederic Jameson in 1997. His chapter is "Technology and the Humanities in the 'Global' Economy."

Kathie L. Olsen is deputy director of the National Science Foundation and former chief scientist of the National Aeronautics and Space Administration (NASA). Her research concerns the neural and genetic mechanisms underlying the development and expression of behavior. Dr. Olsen has published numerous scientific articles and book chapters and has coedited *The Development of Sex Differences and Similarities in Behavior*. A former Brookings Institute Legislative Fellow, Dr. Olsen has received a number of awards for her research, including the International Behavioral Neuroscience Society's Award and the Society for Behavioral Neuro-Endocrinology Award for Outstanding Contributions for Research and Education. Her chapter, cowritten with Dr. Ann B. Carlson, is "Federal Science Policy and University Research Agendas."

Timothy G. Reeves is a specialist in sustainable agricultural production, director of Timothy G. Reeves and Associates Ltd. in Australia, and the former Director General of the International Maize and Wheat Improvement Center near Mexico City, known around the world by its Spanish acronym, CIMMYT (Centro Internacional para Mejoramiento de Maiz y Trigo). As a nonprofit organization, CIMMYT has a goal of ending world hunger in a way that preserves land resources and natural resources while enhancing people's lives. Its traditional research focus has been maize and wheat, and it cooperates in research partnerships worldwide. Dr. Reeves was a member of the Millennium Project's Hunger Task Force in 2003 and served previously as chief scientist of the Department of Agriculture in the State of Victoria, Australia, and as a professor at the University of Adelaide in South Australia. Dr. Reeves has received a number of honors and awards, including an honorary professorship of the Chinese Academy of Sciences. Professor Reeves has been honored by countries including the Philippines, India, and Nepal. Dr. Reeves's chapter, co-written by Kelly Cassaday, is "Global Public Goods for Poor Farmers: Myth or Reality?"

James D. Savage is currently the executive assistant to the president of the University for Federal Relations and a professor in the Department of Politics at the University of Virginia. He is a prolific author, including his recent books, *Funding Science in America: Congress, Universities, and the Politics of the Academic Pork Barrel*, published by Cambridge University Press, and *Balanced Budgets and American Politics*, published by Cornell University Press. His chapter is "The Ethical Challenges of the Academic Pork Barrel."

Larry Smarr has long been a pioneer in the prototyping of a national information infrastructure to support academic research, governmental functions, and industrial competitiveness. In 1983, Dr. Smarr initiated the first proposal to the National Science Foundation recommending the development of a national supercomputer center. This resulted in the creation of the National Center for Supercomputing Applications (NCSA) at the University of Illinois at Urbana-Champaign (UIUC), where he served as director until March 2000. During his years at the NCSA, Dr. Smarr worked very closely with industry, including such companies as SGI, Hewlett-Packard, IBM, Compaq, Sun Microsystems, Intel, Microsoft, Ameritech, AT&T, Qwest, MCI, Cisco, and EMC, in creating today's information infrastructure. In 1997, Dr. Smarr became the director of the National Computational Science Alliance, composed of over fifty universities, government labs, and corporations linked to the NCSA in a national-scale virtual enterprise to prototype the information infrastructure of the twenty-first century. In 1990, he received the Franklin Institute's Delmer S. Fahrney Medal for Leadership in Science or Technology. In 2000, Dr. Smarr became a professor of computer science and engineering in the Jacobs School of Engineering at the University of California at San Diego. He continues to be an active member of a number of high-level government committees such as the President's Information Technology Advisory Committee and the Advisory Committee to the Director, National Institutes of Health. Dr. Smarr is the author of many scholarly articles and the coauthor, with William Kaufmann III, of the book *Supercomputing and the Transformation of Science*. His chapter is "Back to the Future: The Increasing Importance of the States in Setting the Research Agenda."

M. S. Swaminathan, the winner of the 1987 World Food Prize and the 1999 Indira Ghandi Award, is known throughout the world as the "Father of the Green Revolution," an agricultural revolution in India and other Third World countries that radically improved agricultural yields and thereby saved populations from starvation through the introduction of genetically superior grain varieties. This work transformed India from a country with a severe food deficit into a nation that nearly doubled its total agricultural output from 12 million metric tons to 23 million metric tons in just four cropping seasons—an incredible achievement by any standard. Today, Dr. Swaminathan is active in the leadership of the Millennium Project's Hunger Task Force, the International Union for the Conservation of Nature and Natural Resources, the Indian Agricultural Research Institute, and the Indian Council for Agricultural Research. He is the secretary of the Ministry of Agriculture and Irrigation in India, leader of the International Rice Research Institute in the Philippines, chairman of the U.N. World Food Congress in Rome, and chairman of the M. S. Swaminathan Research Foundation. Cited by *Time* magazine as one of the twenty most influential Asians in the twentieth century, Dr. Swaminathan has over thirty-eight honorary doctorates from institutions on three continents. His chapter is entitled "Science and Sustainable Food Security."